ELECTROGRAVITICS II

Validating Reports on a New Propulsion Methodology

Thomas Valone, PhD, PE

Integrity Research Institute
Washington DC
A nonprofit 501(c)3 organization

for T. Townsend Brown

Published by

Integrity Research Institute
5020 Sunnyside Avenue, Suite 209
Beltsville MD 20705
202-452-7674, 301-220-0440

www.IntegrityResearchInstitute.org

Electrogravitics II
Thomas F. Valone, PhD, PE

Cover illustration (right half) © 2000, Mark McCandlish
B-2 Photos courtesy of Bobbi Garcia and Northrop Grumman
Thanks to Thomas Bahder, Takaaki Musha, Jeane Manning, Mark
McCandlish, Dr. B, Dr. Steven Greer, Richard Boylan and
T. Townsend Brown for their contributions

First Edition, December, 2004
Second Edition, July, 2005
Third Edition, January, 2008

ISBN 978-0-9641070-9-0

See last page to request your FREE companion EGII CD

TABLE OF **CONTENTS**

3

FOREWORD

Steven Greer, M.D.

New Energy Solutions and Implications
For The National Security and the Environment

The ultimate national security issue is intimately linked to the pressing environmental crisis facing the world today: The question of whether humanity can continue as a technologically advanced civilization.

Fossil fuels and the internal combustion engine are non-sustainable both environmentally and economically – and a replacement for both already exists. The question is not whether we will transition to a new post-fossil fuel economy, but when and how. The environmental, economic, geopolitical, national security, and military issues related to this matter are profound and inextricably linked to one another.

The disclosure of such new energy technologies will have far-reaching implications for every aspect of human society and the time has come to prepare for such an event. For if such technologies were announced today, it would take at least 10-20 years for their widespread application to be effected. This is approximately how much time we have before global economic chaos begins due to demand far exceeding the supply of oil and environmental decay becomes exponential and catastrophic.

We have found that the technologies to replace fossil fuel usage already exist and need to be exploited and applied immediately to avert a serious global economic, geopolitical, and environmental crisis in the not-so-distant future.

In summary, these technologies fall into the following broad categories:

- Quantum vacuum/zero point field energy access systems and related advances in electromagnetic theory and applications
- Electrogravitic and magnetogravitic energy and propulsion
- Room temperature nuclear effects
- Electrochemical and related advances to internal combustion systems which achieve near zero emissions and very high efficiency

A number of practical applications using such technologies have been developed over the past several decades but such breakthroughs have been either ignored due to their unconventional nature or have been classified and suppressed due to national security, military interests, and 'special' interests.

Let us be clear: the question is not whether such systems exist and can be viable replacements for fossil fuels. The question is whether we have the courage to allow such a transformation in world society to occur.

Such technologies – especially those which bypass the need to use an external fuel source such as oil or coal – would have obvious and beneficial effects for humanity. Since these technologies do not require an expensive source of fuel but instead use existing quantum space energy, a revolution in the world's economic and social order would result.

Implications of Applying Such Technologies

These implications include:

- The removal of all sources of air pollution related to energy generation, including electric power plants, cars, trucks, aircraft and manufacturing.
- The ability to 'scrub' to near zero effluent all manufacturing processes since the energy per se required for same would have no cost related to fuel consumption. This would allow the full application of technologies which remove effluent smokestacks, solid waste, and waterways since current applications are generally restricted by their energy costs and the fact that such energy consumption – being fossil fuel-based – soon reaches the point of diminishing returns environmentally.
- The practical achievement of an environmentally near-zero impact yet high tech civilization on earth, thus assuring the long-term sustainability of human civilization.
- Trillions of dollars now spent on electric power generation, gas, oil, coal and nuclear power would be freed to be spent on more productive and environmentally neutral endeavors by both individuals and society as a whole.

- Underdeveloped regions of the earth would be lifted out of poverty and into a high technology world in about a generation but without the associated infrastructure costs and environmental impact related to traditional energy generation and propulsion. Since these new systems generate energy from the ambient quantum energy state, trillion dollar infrastructure investments in centralized power generation and distribution would be eliminated. Remote villages and towns would have the ability to generate energy for manufacturing, electrification, water purification etc. without purchasing fuels or building massive transmission lines and central power grids.

- Near total recycling of resources and materials would be possible since the energy costs for doing so – now the main obstacle would be brought down to a trivial level.

- The vast disparity between rich and poor nations would quickly disappear and it much of the zero-sum-game mentality which is at the root of so much social, political, and international unrest. In a world of abundant and inexpensive energy, many of the pressures which have led to a cycle of poverty, exploitation, resentment, and violence would be removed from the social dynamic. While ideological, cultural and religious differences would persist, the raw economic disparity and struggle would be removed from the equation fairly quickly.

- Surface roads – and therefore most road building – will be unnecessary as electrogravitic antigravity energy and propulsion systems replace current surface transportation systems.

- The world economy would expand dramatically and those advanced economies such as in the US and Europe would benefit tremendously as global trade, development and high technology energy and propulsion devices are demanded around the world. Such a global energy revolution would create an expanding world economy which would make the current computer and Internet economy look like a rounding error. This really would be the tide which would lift all ships.

- Long term, society would evolve to a psychology of abundance, which would redound to the benefit of humanity as a whole, a peaceful civilization and a society focused increasingly on creative pursuits rather than destructive and violent endeavors.

7

Lest all of this sound like a pipe dream, keep in mind that such technological advances are not only possible, *but they already exist*. What is lacking is the collective will, creativity and courage to see that they are applied wisely. And therein lies the problem.

As an emergency and trauma doctor, I know that everything can be used for good or for ill. A knife can butter your bread – or cut your throat. Every technology can have beneficial as well as harmful applications.

The latter partially explains the serious national security and military concerns with such technologies. For many decades, these advances in energy and propulsion technologies have been acquired, suppressed and classified by certain interests who have viewed them as a threat to our security from both an economic and military perspective. In the short term, these concerns have been well-founded: Why rock the global economic boat by allowing technologies out which would, effectively, terminate the multimillion dollar oil, gas, coal, internal combustion engine and related transportation sectors of the economy? And which could also unleash such technologies on an unstable and dangerous world where the weapons applications for such technological breakthroughs would be a certainty? In the light of this, the status quo looks good.

But only for the short term. In fact, such national security and military policies – fed by huge special interests in obvious industries and nations – have exacerbated global geopolitical tensions by impoverishing much of the world, worsening the zero-sum-game mind set of the rich versus poor nations and brought us to a world energy emergency and a pending environmental crisis. And now we have very little time to fix the situation. Such thinking must be relegated to the past.

For what can be a greater threat to the national security than the specter of a collapse of our entire civilization from a lack of energy and global chaos as every nation fights for its share of a limited resource? Due to the long lead-time needed to transform the current industrial infrastructure away from fossil fuels, we are facing a national security emergency which almost nobody is talking about. This is dangerous.

It has also created a serious constitutional crisis in the US and other countries where non-representative entities and super-secret projects within compartmented military and corporate areas have begun to set national and international policy on this and related matters – all outside the arena of public debate, and mostly without informed consent from Congress or the President.

Indeed this crisis is undermining democracy in the US and elsewhere. I have had tlle unenviable task of personally briefing senior political, military, and intelligence officials in the US and Europe on this and related matters. These officials have been denied access to information compartmented within certain projects which are, frankly, unacknowledged areas (*so-called 'black' projects*). Such officials include members of the House and Senate, President Clinton's first Director of Central Intelligence, the head of the DIA, senior Joint Staff officials and others.

Usually, the officials have little to no information on such projects and technologies – and are told either nothing or that they do not have a 'need to know' if they specifically inquire.

This presents then another problem: *these technologies will not be suppressed forever*. For example, our group is planning a near term disclosure of such technologies and we will not be silenced. At the time of such a disclosure, will the US government be prepared? It would behoove the US government and others to be informed and have a plan for transitioning our society from fossil fuels to these new energy and propulsion systems.

Indeed, the great danger is ignorance by our leaders of these scientific breakthroughs – and ignorance of how to manage their disclosure. The advanced countries of the world must be prepared to put systems in place to assure the exclusive peaceful use of such energy and propulsion advances. Economic and industrial interests should be prepared so that those aspects of our economy which will be adversely affected (commodities, oil, gas, coal, public utilities, engine manufacturing, etc.) can be cushioned from sudden reversals and be economically 'hedged' by investing in and supporting the new energy infrastructure.

New Energy Solutions

A creative view of the future – not fear and suppression of such technologies – is required. And it is needed immediately. If we wait 10-20 more years, it will be too late to make the needed changes before world oil shortages, exorbitant costs and geopolitical competition for resources causes a melt-down in the world's economy and political structures.

All systems tend towards homeostasis. The status quo is comfortable and secure. Change is frightening. But in this case, the most dangerous course for the national security is inaction. We must be prepared for the coming convulsions related to energy shortages,

9

spiraling costs and economic disruption. The best preparation would be a replacement for oil and related fossil fuels. And we have it. But disclosing these new energy systems carries its own set of benefits, risks and challenges. The US government and the Congress must be prepared to wisely manage this great challenge.

Recommendations for Congress:

- Thoroughly investigate these new technologies both from current civilian sources as well as compartmented projects within military, intelligence and corporate contracting areas.
- Authorize the declassification and release of information held within compartmented projects related to this subject.
- Specifically prohibit the seizing or suppression of such technologies.
- Authorize substantial funding for basic research and development by civilian scientists and technologists into these areas.
- Develop plans for dealing with disclosing such technologies and for the transition to a non-fossil fuel economy. These plans should include: military and national security planning; strategic economic planning and preparation; private sector support and cooperation; geopolitical planning, especially as it pertains to OPEC countries and regions whose economies are very dependent on oil exports and the price of oil; international cooperation and security; among others.

I personally stand ready to assist the Congress in any way possible to facilitate our use of these new energy sources. Having dealt with this and related sensitive matters for over 10 years, I can recommend a number of individuals who can be subpoenaed to provide testimony on such technologies, as well as people who have information on Unacknowledged Special Access Projects (USAPs) within covert government operations which are already dealing with these issues.

If we face these challenges with courage and with wisdom together, we can secure for our children a new and sustainable world, free of poverty and environmental destruction. We will be up to this challenge because we must be.

Steven Greer, M.D.
Crozet, Virginia

SCIENCE SECTION

J. L. Naudin's latest electric field gradient shaping, asymmetric capacitor lifters from his website www.jlnlabs.org

What is Electrogravitics and Has It Been Validated?

Thomas Valone, PhD, PE

This book offers an updated viewpoint on the confusing and often misinterpreted concept of electrogravitics or electrogravity, compared to electrokinetics. It is now time to set the record straight for the sake of all of the researchers who have sought to learn the truth behind a propulsion mystery spanning almost a century. It is helpful if the reader has already familiar with the first volume, *Electrogravitics Systems: A New Propulsion Methodology* "Volume I", which has been in print for over ten years. However, Volume II both predates and postdates the first volume, thus giving a wider historical perspective.

What is Electrogravitics

When asked, "What is electrogravitics?" a qualified answer is *"electricity used to create a force that depends upon an object's mass, even as gravity does."* This is the answer that I believer should still be used to identify true electrogravitics, which also involves the object's mass in the force, often with a dielectric. This is also what the "Biefeld-Brown effect" of Brown's first patent #300,311 describes. However, we have seen T. Townsend Brown and his patents evolve over time which Tom Bahder emphasizes. Later on, Brown refers to **"electrokinetics,"** which requires asymmetric capacitors and pulsed current to amplify the force. Therefore, Bahder's article discusses the lightweight effects of "lifters" and the ion mobility theory found to explain them. Note: *electrogravitics includes electrokinetics.*

To put things in perspective, the article "How I Control Gravitation," published in 1929 by Brown,[1] presents an electrogravitics-validating discovery about very heavy metal objects (44 lbs. each) separated by an insulator, charged up to high voltages. T.T. Brown also expresses an experimental formula in words which tell us what he found was directly contributing to the *unidirectional force* (UDF) which he discovered, moving the system of masses toward the positive charge. He seems to imply that the equation for his electrogravitic force might be $F \approx Vm_1m_2/r^2$. But electrokinetics and electrogravitics also seem to be governed by another equation (Eq.1).

---

[1] Reprinted on p. 71 of this book.

Zinsser Effect versus the Biefeld-Brown Effect

There is another very similar invention which has comparable experiments that also involve electrogravity. It is the discovery of "gravitational anisotropy" by Rudolf G. Zinsser from Germany. I met with Zinsser twice in the early 1980's and corresponded with him subsequently regarding his invention. He presented his experimental results at the Gravity Field Conference in Hanover in 1980, and also at the First International Symposium of Non-Conventional Energy Technology in Toronto in 1981.[2] For years afterwards, all of the scientists who knew of Zinsser's work regarded his invention as a unique phenomenon, not able to be classified with any other discovery. However, upon reading Brown's 1929 article on gravitation referred to above, I find striking similarities.

Zinsser's discovery is detailed in *The Zinsser Effect* book by this author.[3] To summarize his life's work, Zinsser discovered that if he connected his patented pulse generator to two conductive metal plates immersed in water, he could induce a sustained force that lasted even after the pulse generator was turned off. The pulses lasted for only a few nanoseconds each.[4] Zinsser called this input "a kinetobaric driving impulse." Furthermore, he points out in the Specifications and Enumerations section, reprinted in my book, that the high dielectric constant of water (about 80) is desirable and that a solid dielectric is possible. Dr. Peschka calculated that Zinsser's invention produced 6 Ns/Ws or 6 N/W.[5] This figure is *twenty times* the force per energy input of the Inertial Impulse Engine of Roy Thornson, which has been estimated to produce 0.32 N/W.[6] By comparison, it is important to realize that any production of force today is <u>extremely inefficient</u>, as seen by the fact that a DC-9 jet engine produces only 0.016 N/W or 3 lb/hp (fossil-fuel-powered land and air vehicles are even worse.)

---

[2] Zinsser, R.G. "Mechanical Energy from Anisotropic Gravitational Fields" First Int'l Symp. on Non-Conventional Energy Tech. (FISONCET), Toronto, 1981. Proceedings available from PACE, 100 Bronson Ave #1001, Ottawa, Ontario K1R 6G8

[3] Valone, Thomas *The Zinsser Effect: Cumulative Electrogravity Invention of Rudolf G. Zinsser*, Integrity Research Institute, 2005, 130 pages, IRI #701

[4] Cravens, D.L. "Electric Propulsion/Antigravity" *Electric Spacecraft Journal*, Issue 13, 1994, p. 30

[5] Peschka, W., "Kinetobaric Effect as Possible Basis for a New Propulsion Principle," *Raumfahrt-Forschung*, Feb, 1974. Translated version appears in *Infinite Energy*, Issue 22, 1998, p. 52 and *The Zinsser Effect*.

[6] Valone, Thomas, "Inertial Propulsion: Concept and Experiment, Part 1" Proc. of Inter. Energy Conver. Eng. Conf., 1993, See IRI Report #608.

Let's now compare the Zinsser Effect with the Biefeld-Brown Effect, looking at the details. Brown reports in his 1929 article that there are effects on plants and animals, as well as effects from the sun, moon and even slightly from some of the planetary positions. Zinsser also reports beneficial effects on plants and humans, including what he called "bacteriostasis and cytostasis."[7] Brown also refers to the "endogravitic" and "exogravitic" times that were representative of the charging and discharging times. Once the gravitator was charged, depending upon "its gravitic capacity" any further electrical input had no effect. This is the same phenomenon that Zinsser witnessed and both agree that the *pulsed voltage generation* was the main part of the electrogravitic effect.[8]

Both Zinsser and Brown worked with dielectrics and capacitor plate transducers to produce the electrogravitic force. Both refer to a high dielectric constant material in between their capacitor plates as the preferred type to best insulate the charge. However, Zinsser never experimented with different dielectrics nor higher voltage to increase his force production. This was always a source of frustration for him but he wanted to keep working with water as his dielectric.

Electrically Charged Torque Pendulum of Erwin Saxl

Brown particularly worked with a torque (torsion) pendulum arrangement to measure the force production. He also refers the planetary effects being most pronounced *when aligned with the gravitator* instead of perpendicular to it. He compares these results to Saxl and Allen, who worked with an electrically charged torque pendulum.[9] Dr. Erwin Saxl used high voltage in the range of +/- 5000 volts on his very massive torque pendulum.[10] The changes in period of oscillation measurements with solar or lunar eclipses, showed great sensitivity to the shielding effects of gravity during an alignment of astronomical bodies, helping to corroborate Brown's observation in his

---

[7] See "Pulsed Electromagnetic Field Health Effects" IRI Report #418 and *Bioelectromagnetic Healing* book #414 by this author, which explain the beneficial therapy which PEMFs produce on biological cells.

[8] Mark McCandlish's Testimony (p. 131) shows that the Air Force took note in that the electrogravitic demonstration craft shown at Norton AFB in 1988 had a rotating distributor for electrically pulsing sections of multiply-layered dielectric and metal plate pie-shaped sections with high voltage discharges.

[9] See Saxl patent #3,357,253 "Device and Method for Measuring Gravitational and Other Forces" which uses +/- 5000 volts.

[10] Saxl, E.J., "An Electrically Charged Torque Pendulum" *Nature*, July 11, 1964, p. 136

1929 article. The pendulum Saxl used was over 100 kilograms in mass.[11] Most interesting were the "unexpected phenomena" which Saxl reported in his 1964 *Nature* article (see footnote 10). The positively charge pendulum had the longest period of oscillation compared to the negatively charged or grounded pendulum. Dirunal and seasonal variations were found in the effect of voltage on the pendulum, with the most pronounced occurring during a solar or lunar eclipse. In my opinion, this demonstrates the basic principles of electrogravitics: high voltage and mass together will cause unbalanced forces to occur. In this case, the electrogravitic interaction was measurable by oscillating the mass of a charged torque pendulum (producing current) whose period is normally proportional to its mass.

Electrogravitic Woodward-Nordtvedt Effect[12]

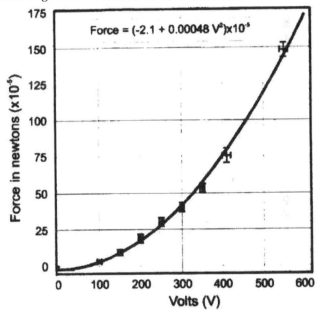

Fig. 1 Force (10^{-5} N = dynes) output vs. capacitor voltage (V) input of a Woodward force transducer "flux capacitor"

[11] Saxl & Allen, "Observations with a Massive Electrified Torsion Pendulum: Gravity Measurements During Eclipse," IRI Report #702.(Note: 2.2 lb = 1 kg)

[12] Graph of Fig. 1 from Woodward and Mahood, "Mach's Principle, Mass Fluctuations, and Rapid Spacetime Transport," California State University

Referring to mass, it is sometimes not clear whether gravitational mass or inertial mass is being affected. The possibility of altering the equivalence principle (which equates the two), has been pursued diligently by Dr. James Woodward, whose patents can be reviewed in the Patent Section of this book. His prediction, based on Sciama's formulation of Mach's Principle in the framework of general relativity, is that "in the presence of *energy flow*, the inertial mass of an object may undergo sizable variations, changing as the 2nd time derivative of the energy."[13] Woodward, however, indicates that it is the "active gravitational mass" which is being affected but the equivalence principle causes both "passive" inertial and gravitational masses to fluctuate.[14] With barium titanate dielectric between disk capacitors. a 3 kV signal was applied in the experiments of Woodward and Cramer resulting in symmetrical mass fluctuations on the order of centigrams.[15] Cramer actually uses the phrase "Woodward effect" in his AIAA paper, though it is well-known that Nordtvedt was the first to predict noticeable mass shifts in accelerated objects.[16]

The interesting observation which can be made, in light of previous sections, is that Woodward's experimental apparatus *resembles a combination of Saxl's torsion pendulum and Brown's electrogravitic dielectric capacitors.* The differences arise in the precise timing of the pulsed power generation and with input voltage. Recently, 0.01 μF capacitors (Model KD 1653) are being used, in the 50 kHz range (lower than Zinsser's 100 kHz) with the voltage still below 3 kV. Significantly, the thrust or unidirectional force (UDF) is exponential, depending on the square of the applied voltage.[17] However, the micronewton level of force that is produced is actually the same order of magnitude which Zinsser produced, who reported his results in dynes (1 dyne = 10^{-5} Newtons).[18] Zinsser had *activators* with masses between 200 g and 500 g and force production of "100 dynes to over

---

Fullerton, Fullerton CA 92634

[13] Cramer et al., "Tests of Mach's Principle with a Mechanical Oscillator" AIAA-2001-3908 email: cramer@phys.washington.edu

[14] Woodward, James F. "A New Experimental Approach to Mach's Principle and Relativistic Gravitation, *Found. of Phys. Letters*, V. 3, No. 5, 1990, p. 497

[15] Compare Fig. 1 graph to Brown's ONR graph on P.117 of Volume I

[16] Nordtvedt, K. *Inter. Journal of Theoretical Physics*, V. 27, 1988, p. 1395

[17] Mahood, Thomas "Propellantless Propulsion: Recent Experimental Results Exploiting Transient Mass Modification" Proc. of STAIF, 1999, CP458, p. 1014 (Also see Mahood Master's Thesis www.serve.com/mahood/thesis.pdf)

[18] For comparison, 1 Newton = 0.225 pounds – Ed. note

one pound."[19] Recently, Woodward has been referring to his transducers as "flux capacitors" (like the movie, *Back to the Future*).[20]

Jefimenko's Electrokinetics Explains Electrogravitics

Known for his extensive work with atmospheric electricity, electrostatic motors and electrets, Dr. Oleg Jefimenko deserves significant credit for presenting a valuable theory of the *electrokinetic field*, as he calls it.[21] A W.V. University professor and physics purist at heart, he describes this field as the <u>dragging force</u> that electrons exert <u>on neighboring electric charges</u>. He identifies the *electrokinetic field* by the vector $\mathbf{E}_k$ where

$$\mathbf{E}_k = -\frac{1}{4\pi\varepsilon_o c^2} \int \frac{1}{r}\left[\frac{\partial \mathbf{J}}{\partial t}\right] dv' \qquad (1)$$

It is one of three terms for the electric field in terms of current and charge density. Equations like $\mathbf{F} = q\mathbf{E}$ also apply for calculating force.

The significance of $\mathbf{E}_k$, as seen in Eq. 1, is that the electrokinetic field simply the third term of the classical equation for <u>the electric field</u>:

$$\mathbf{E} = \frac{1}{4\pi\varepsilon_o} \int \left\{\frac{\rho}{r^2} + \frac{1}{rc}\frac{\partial\rho}{\partial t}\right\} \mathbf{r}\, dv' + \mathbf{E}_k \qquad (1a)$$

This three-term equation is a "causal" equation, according to Jefimenko, because it links the electric field $\mathbf{E}$ back the electric charge and its motion (current) which induces it. This is the essence of electromagnetic induction, *as Maxwell intended*, which is measured by, not caused by, a changing magnetic field. The second electric field term, designated as the electrokinetic field, is directed along the current direction or parallel to it. It also exists only as long as the current is changing in time. Lenz' Law is also built into the minus sign. <u>Parallel conductors</u> will produce the strongest induced current.

By examining the vector potential $\mathbf{A}$ equation which depends upon the current density $\mathbf{J}$, he finds that $\mathbf{E}_k$ can be expressed as the time derivative of $\mathbf{A}$, which leads to

$$\mathbf{A} = -\int \mathbf{E}_k dt + const. \qquad (2)$$

---

[19] Zinsser, FISONCET, Toronto, 1981, p. 298

[20] Woodward, James "Flux Capacitors and the Origin of Inertia" *Foundations of Physics*, V. 34, 2004, p. 1475. Also see "Tweaking Flux Capacitors" *Proc. of STAIF*, 2005

[21] Jefimenko, Oleg *Causality, Electromagnetic Induction and Gravitation*,

The significance of Eq. 2 is that the magnetic vector potential is seen to be created by the time integral which amounts to an *electrokinetic impulse* "produced by this current at that point <u>when the current is switched on</u>" according to Jefimenko.[22] Of course, a time-varying sinusoidal current will also qualify for production of an electrokinetic field and the vector potential. An important consequence of Eq. 1 is that *the faster the rates of change of current, the larger will be the electrokinetic force.* Therefore, high voltage pulsed inputs are favored.

However, its significance is much more general. "This field can exist anywhere in space and can <u>manifest itself as a pure force</u> by its action on free electric charges." All that is required for a measurable force *from a single conductor* is that the change in current density (time derivative) happens very fast, to overcome the c^2 in the denominator.

The electrogravitics experiments of Brown and Zinsser involve a

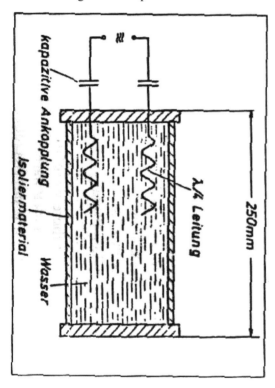

Fig. 2
Sample capacitor probe used by Zinsser. Notice the quarter λ/4 wavelength electrodes which indicate a resonant circuit design.

Electret Scientific Co., POB 4132, Star City, WV 26504, p. 29
[22] Jefimenko, p. 31

dielectric medium for greater efficacy and charge density. The electrokinetic force on the electric charges (electrons) of the dielectric, according to Eq. 1, is in the *opposite direction of the increasing positive current* (taking into account the minus sign). For parallel plate capacitors, Jefimenko explains that *the strongest induced field is produced between the plates* and so another equation evolves.

Electrokinetic Force Predicts Electrogravitic Direction
Can Jefimenko's electrokinetic force predict the correct *direction* of the electrogravitic force seen in the Zinsser, Brown, Woodward as well as the yet-to-be-discussed Campbell, Serrano, and Norton AFB craft demonstrations?
1) Starting with <u>Zinsser's probe diagram</u> (Fig. 2) from Peschka's article, it is purposely put on its end for reasons that will become obvious. Compare it with an equivalent parallel plate capacitor (the plates are x distance apart) from Jefimenko's book:[23]

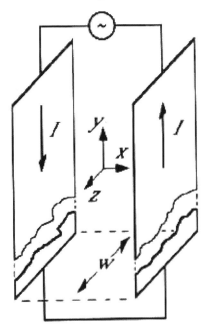

Fig. 3
Calculation of Jefimenko's electrokinetic force in the space between two current-carrying plates. X is the space between the plates. W is the width of the plates.

---

[23] Jefimenko, p. 47

19

We note that the current is presumed to be the same in each plate but in opposite directions because it is alternating. Using Eq. 2, Jefimenko calculates the electrokinetic field, for the AC parallel plate capacitor with current going in opposite directions, as

$$\mathbf{E}_k = -\mu_o \frac{\partial I}{\partial t} \frac{x}{w} \mathbf{j} \qquad (3)$$

Of course, in vector calculus, $\mathbf{i}$, $\mathbf{j}$, and $\mathbf{k}$ are the unit vectors for the x, y, z axis directions seen in Fig. 3, respectively. It is clearly seen that the y-axis points upward in Fig. 3 and so with the minus sign of Eq. 3, the electrokinetic force for the AC parallel plate capacitor *will point downward*. Since Zinsser had his torsion balance on display in Toronto in 1981, I was privileged to verify the direction of the force that is created with his quarter-wave plates oriented as they are in Fig. 2. The torsion balance is built so that the capacitor probe can only be deflected *downward* from the horizontal. *The electrokinetic force is in the same direction.*

    2) Looking at <u>Brown's electrogravitic force direction</u> from the Fig. 1

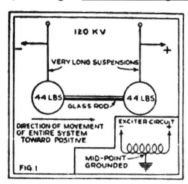

A SIMPLE TYPE OF GRAVITATOR IS SHOWN IN THE ABOVE ILLUSTRATION.

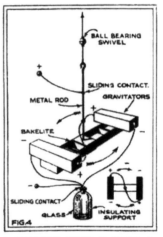

A GRAVITATOR ROTOR IS SIMPLY AN ASSEMBLY OF UNITS SO MADE THAT ROTATION RESULTS UNTIL THE IMPULSE IS EXHAUSTED.

in his 1929 article "How I Control Gravitation," we see that the positive lead is on the right side of the picture. Also, the arrow below *points to the right* with the caption, "Direction of movement of entire system toward positive." Examining the electrokinetic force of Eq. 1 in this article, we note that the increasing positive current comes in by convention in the positive lead and points to the left. Therefore, considering the minus sign, the direction of the electrokinetic force will be *to the right*. Checking with

Fig. 4 of the 1929 Brown article, the same *confirmation of induced electrokinetic force direction*.[24] Thus, with Zinsser's and Brown's gravitators, *the electrokinetic theory provides a useful explanation and it is accurate for prediction of the resulting force direction.*

It is also worthwhile noting that Brown also indicates in that article,

"when the direct current with high voltage (75 – 300 kilovolts) is applied, the gravitator swings up the arc ... but it does not remain there. The pendulum then gradually returns to the vertical or starting position, even while the potential is maintained...Less than five seconds is required for the test pendulum to reach the maximum amplitude of the swing, but from thirty to eighty seconds are required for it to return to zero."

This phenomena is *remarkably the same type of response that Zinsser recorded* with his experimental probes. Jefimenko's theory helps explain the rapid response, since the change of current happens in the beginning. However, the slow discharge in both experiments (which Zinsser called a "storage effect") needs more consideration. Considering the electrokinetic force of Eq. 3 and the +/- derivative, we know that the slow draining of a charged capacitor, most clearly seen

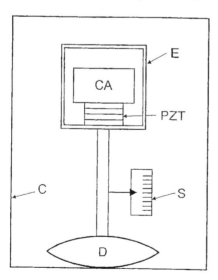

Fig. 4
Woodward's #6,098,924 patented impulse engine, also called a flux capacitor. The PZT provides nanometer-sized movements that are timed to an AC signal input. A torsion balance has been used with a pair of force transducers in other designs.

---

[24] Brown's second patent #2,949,550 (see Patent Section: two electrokinetic saucers on a maypole) has movement toward the positive charge, so the same

in Fig. 1 of Brown's 1929 article, will produce a decreasing current out of the + terminal (to the right) and in Eq. 3, this means the derivative is negative. Therefore, the slow draining of current will produce a weakening electrokinetic force but *in the same direction as before!* The force will thus sustain itself to the right during discharge.

3) It is very likely that the electrokinetic theory will also predict the direction of Woodward's UDF but instantaneous analysis needs to be made to compare current direction into the commercial disk capacitors and the electrokinetic force on the dielectric charges. In every electrogravitics or electrokinetics case, it can be argued, the "neighboring charges" to a capacitor plate will necessarily be those in the dielectric material, which are polarized. The bound electron-lattice interaction *will drag the lattice material with them*, under the influence of the electrokinetic force. If the combination of physical electron acceleration (which also can be regarded as current flow) and the AC signal current flow can be resolved, it may be concluded that an instantaneous electrokinetic force, depending on dI/dt, contributes to the Woodward-Nordtvedt effect.

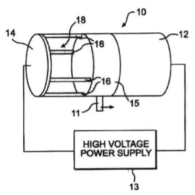

Fig. 5 Capacitor module from Campbell's NASA patent #6,317,310 which creates a thrust force. Disk 14 is copper; Struts 16 are dielectrics; Cylinder 15 is a dielectric; Cylinder 12 is an axial capacitor plate; Support post 11 is also dielectric.

4) The Campbell and Serrano capacitor modules seen in the Patent Section, as well as the Electrogravitic Craft Demonstration unit (Norton AFB), can also be explained with the electrokinetic force, in the same way that the Brown gravitator force was explained in paragraph (2) above. The current flows in one direction through the capacitor-dielectric and the force is produced in the opposite direction. The Norton AFB electrogravitic craft just has bigger plates with radial

---

electrokinetic theory explained above works for both. – Ed note

sections but the current flow still occurs at the center, *across the plates*. The Serrano patent diagram is also very similar in construction and operation.

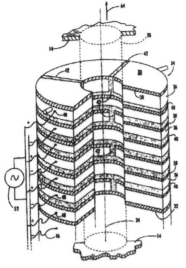

Fig. 6
Capacitor propulsion device with alternating metal and dielectric layers from Serrano's PCT patent WO 00/58623 with upward thrust direction indicated and + and – polarity designated on the side.

Electrokinetic Theory Observations

For parallel plate capacitor impulse probes, like Zinsser, Serrano, Campbell, the Norton AFB craft and both of Brown's models, the electrokinetic field of Eq. 3 provides a working model that seems to predict the *nature and direction of the force* during charging and discharging phases. More detailed information is needed for each example in order to actually calculate the theoretical electrokinetic force and compare it with experiment. We note that Eq. 3 also does not suffer the handicap of Eq. 1 since no c^2 term occurs in the denominator. Therefore, it can be concluded that AC fields operating on parallel plate capacitors should create significantly larger electrogravitic forces than other geometries with the same dI/dt. However, the current I is usually designated as $I_o\sin(\omega t)$ and its derivative is a sinusoid as well. Therefore, a detailed analysis is needed for each specific circuit and signal to determine the outcome.

Eq. 3 also suggests a *possible enhancement* of the force if a permeable dielectric (magnetizable) is used. Then, the value for μ of the material would normally be substituted for μ_o.[25]

---

[25] Einstein and Laub, *Annalen der Physik*, V. 26, 1908, p.533 and p. 541 – two articles on the subject of a moving capacitor with a "dielectric body of considerable permeability." Specific equations are derived predicting the

A further observation of both Eq. 1 and Eq. 3 is that very fast changes in current, such as *a current surge or spark discharge* has to produce the most dynamic electrokinetic force, since dI/dt will be very large.[26] The declining current surge, or the negatively sloped dI/dt however, should create an opposing force until the current reverses direction. *Creative waveshaping seems to be the answer* to this

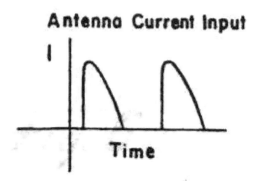

Antenna Current Input

Fig. 7 The ideal electrokinetic force current waveform is found in Schlicher propulsion patent #5,142,861

obvious dilemma. Fortunately, a few similar inventions use pulse power electric current generators to create propulsion. The Taylor patent #5,197,279 "Electromagnetic Energy Propulsion Engine" uses huge currents to produce magnetic field repulsion. The Schlicher patent #5,142,861 "Nonlinear Electromagnetic Propulsion System and Method" predicts hundreds of pounds of thrust with tens of kiloamperes input. The Schlicher antenna current input is a rectified current surge produced with an SCR-triggered DC power source (see Fig. 7). The resulting waveform has a very steep leading edge but a *slowly declining trailing edge*, which should also be desirable for the electrokinetic force effect.[27]

Another observation that should be mentioned is that this electrokinetic force theory does not include the mass contribution to

resulting EM fields. Translated articles are reprinted in *The Homopolar Handbook* by this author (p. 122-136). Also see Clark's dielectric homopolar generator patent #6,051,905.

[26] Commentary to Eq. 2 states an electrokinetic impulse is produced when the "current is switched on," which implies a very steep leading edge of the current slope.

[27] See the Taylor and Schlicher patents in the Patent Section. – Ed note

the electrogravitic force which Saxl, Woodward, and Brown's 1929 gravitator emphasize. A contributor to this Volume II anthology, Takaaki Musha offers a derived equation for electrogravitics *that does include a mass term* but not a derivative term. His model is based on the charge displacement or "deformation" of the atom under the influence of a capacitor's 18 kV high voltage field and his experimental results are encouraging. He also includes a reference to Ning Li and her *gravitoelectric theory*.[28]

A final concern, which may arise from the very nature of the electrokinetic force description, is the difficulty of conceptualizing or simply accepting the possibility of an *unbalanced force creation pushing against space*. This author has wrestled with this problem in other arenas for years. Three examples include (1) the homopolar generator which creates *back torque* that ironically, <u>pushes against space</u> to implement the Lorentz force to slow down the current-generating spinning disk.[29] Secondly (2), there is the intriguing *spatial angular momentum discovery* by Graham and Lahoz.[30] They have shown, reminiscent of Feynman's "disk paradox," that the vacuum is the seat of Newton's third law. A torsion balance is their chosen apparatus as well to demonstrate the pure reaction force with induction fields. Their reference to Einstein and Laub's papers cites the time derivative of the Poynting vector $S = E \times H$ integrated over all space to preserve Newton's third law. Graham and Lahoz predict that *magnetic flywheels with electrets* will circulate energy to <u>push against space</u> (see Footnote 22). Lastly, for (3), the Taylor and Schlicher inventions <u>push against space</u> with an unbalanced force that is electromagnetic in origin.

Historical Electrogravitics

In the Historical Section, gravity articles like the *NY Herald-Tribune* series and *Interavia* were some of the last few public pronouncements of the progress of this research.[31] They were published in 1955 and

---

[28] Ning Li was the Chair of the 2003 Gravitational Wave Conference. The CD Proceedings of the papers is available from Integrity Research Institute.

[29] Valone, Thomas, *The Homopolar Handbook: A Definitive Guide to Faraday Disk and N-Machine Technologies*, Integrity Research Institute, Third Edition, 2001

[30] Graham and Lahoz, "Observation of Static Electromagnetic Angular Momentum in vacuo" *Nature*, V. 285, May 15, 1980, p. 129

[31] See also "The Flying Saucer" by Mason Rose, PhD, *Science and Invention*, Aug. 1929 and *Psychic Observer*, Vol. XXXVII, No.1

1956 respectively, at the same time when the British Aviation Studies reports spanning 1954 – 56 were published (see Volume I). The aviation industry interest in this science was at an all-time high, mostly spurred on by Brown's gravitator experiments. After all, aircraft are very massive and Brown's theory encourages the use of massive gravitators with high voltage, which we find in the B-2 bomber today.

A fascinating article at the end of this surge in gravity research is the report from 1961 in *Missles and Rockets*,[32] which identifies a 389-page study released by the Office of Technical Services at the US Dept. of Commerce (may possibly be OTS #61-1187). The study however, sadly relates that disagreements among experts were becoming unyielding without more experimental proof.

Today, experimental proof seems to be in abundance. However, the prevailing trend by the government still fails to acknowledge the historical pioneering work of Biefeld and Brown, as well as any small inventor who is successful in this area. Take for example, Hector Serrano, who was interviewed in 1998 by NASA scientist Jonathan Campbell on video, about his electropropulsion invention. Within two years, Campbell started filing for a series of patents "on similar technology" and not referencing T.T. Brown nor Serrano in any of his US patents #6,317,310, #6,411,493, or #6,775,123.[33] This type of behavior by a government representative is unethical and fuels the wide-spread public concern about government motives. Remarkably, it is like history repeating the same treatment that T.T. Brown received from the military upon demonstrating his working model to them.[34]

Eye Witness Testimony of Advanced Electrogravitics

Sincere gratitude is given to Mark McCandlish, who offers us the conclusive perspective of the covert, flat-bottomed saucer hovercraft seen by dozens of invited eye-witnesses at Norton Air Force Base in 1988. When I spoke to Dr. Hal Puthoff about Mark's story, shortly after the famous Disclosure Event[35] at the National Press Club in

[32] Beller, William "Soviet Efforts are Closely Watched" *Missiles and Rockets*, Sept. 11, 1961, p. 27

[33] Young, Kelly "Inventor: NASA stole patent idea" *Florida Today*, Sept. 29, 2002 (entire article is posted on the *Florida Today* website – Ed. note)

[34] See "The Townsend Brown Electro-Gravity Device" File 24-185, A Comprehensive Analysis by the Office of Naval Research, Sept. 15, 1952, - IRI report #612

[35] See the authoritative book by Dr. Steven Greer, *Disclosure: Military and*

2001, he explained to me that he had already performed due diligence on it and checked on each individual to verify the details of the story. Hal told me that he believed the story was true. Since Dr. Puthoff used to work for the CIA for ten years, this was quite an endorsement.

In analyzing the Electrogravitic Craft Demonstration unit (Norton AFB 1988) diagrammed at the end of Mark's testimony, I have compared it to Campbell's and Serrano's patented design. A lot can be learned from studying the intricacies of this advanced design, including the use of a distributor cap style of pulse discharge and multiple symmetric, radial plates with dielectrics in between.

Why Americans should pay twice for the development of 21st century energy and propulsion technology is an issue that several U.S. Congressmen have publicly protested. We pay for the "black project budget" (the difference between the Pentagon's defense budget and its acknowledged expenses) in billions of tax dollars every year.[36] We also are asked to pay for DOE, NASA, AF, Navy, DARPA and other agencies to reinvent the same technologies in an unclassified arena.

Recently, the Deputy Director of the National Reconnaissance Office, for example, told me that it seems to be easier to direct contractors to develop technology *that he knows already exists*, mainly because declassification is very difficult. This is the main reason that we still use World War II technology on land and in space while the environment suffers irreparable harm. My sincere hope is that the validating science contained in *Electrogravitics II* will accelerate the civilian adaptation of this relatively simple propulsion technology.

The scientific articles in the first section of this book show the contrasting opinion that still exists in the assessment of electrogravitics. As *inertial shielding* also is reinvented by civilian scientists, I predict that electrogravitics will become more and more useful. The reason behind my prediction is that any force moving a mass utilizes Newton's Second Law, $F = ma$, which can be very powerful when the inertial mass m is reduced by electrogravitic shielding. Once again, to confirm Dr. Greer's message, such a technological development already exists, as exhibited in the night photos of right-angle turns of covert triangular craft.

Government Witnesses Reveal the Greatest Secretes in Modern History, Crossing Point, 2001. It provides the testimony of each witness who participated in the event, plus many more.
[36] "The Billion Dollar Secret" narrated by defense journalist Nick Cook, aired on TLC in 2000 about the black projects and the money spent on them . He is now filming a follow-up show for 2005.

Force on an Asymmetric Capacitor

Thomas B. Bahder and Chris Fazi

Army Research Laboratory - 2800 Powder Mill Rd - Adelphi, MD 20783

bahder@arl.army.mil

When a high voltage (~30 kV) is applied to a capacitor whose electrodes have different physical dimensions, the capacitor experiences a net force toward the smaller electrode (Biefeld-Brown effect). We have verified this effect by building four capacitors of different shapes. The effect may have applications to vehicle propulsion and dielectric pumps. We review the history of this effect briefly through the history of patents by Thomas Townsend Brown. At present, the physical basis for the Biefeld-Brown effect is not understood. The order of magnitude of the net force on the asymmetric capacitor is estimated assuming two different mechanisms of charge conduction between its electrodes: ballistic ionic wind and ionic drift. The calculations indicate that ionic wind is at least three orders of magnitude too small to explain the magnitude of the observed force on the capacitor. The ionic drift transport assumption leads to the correct order of magnitude for the force, however, it is difficult to see how ionic drift enters into the theory. Finally, we present a detailed thermodynamic treatment of the net force on an asymmetric capacitor. In the future, to understand this effect, a detailed theoretical model must be constructed that takes into account plasma effects: ionization of gas (or air) in the high electric field region, charge transport, and resulting dynamic forces on the electrodes. The next series of experiments should determine whether the effect occurs in vacuum, and a careful study should be carried out to determine the dependence of the observed force on gas pressure, gas species and applied voltage.

1. Introduction

Recently, there is a great deal of interest in the Biefeld-Brown effect: when a high voltage (~30 kV) is applied to the electrodes of an asymmetric capacitor, a net force is observed on the capacitor. By asymmetric, we mean that the physical dimensions of the two electrodes are different, i.e., one electrode is large and the other small. According to the classical Biefeld-Brown effect (see Brown's original 1960, 1962, and 1965 patents cited in Appendix A, and a partial reproduction below), the largest force on the capacitor is in a direction from the negative (larger) electrode toward the positive (smaller) electrode. Today, there are numerous demonstrations of this effect on the Internet in devices called "lifters", which show that the force on the capacitor exceeds its weight [1]. In fact, these experiments indicate that there is a force on the capacitor independent of polarity of applied voltage. In the future, the Biefeld-Brown effect may have application to aircraft or vehicle propulsion, with no moving parts. At the present time, there is no detailed theory to explain this effect, and hence the potential of this effect for applications is unknown. In Section 2 below, we describe the history of the Biefeld-Brown effect. The effect of a net force on an asymmetric capacitor is so surprising, that we carried out preliminary simple experiments at the Army Research Laboratory to verify that the effect is real. The results of these experiments are described in Section 3. Section 4 contains estimates of the force on the capacitor for the case of ballistic ionic wind and drift of carriers across the capacitor's gap between electrodes. In Section 5, we present a detailed thermodynamic treatment of the force on an asymmetric capacitor, assuming that a non-linear dielectric fluid fills the region between capacitor electrodes. Section 6 is a summary and recommendation for future experimental and theoretical work.

2. Biefeld-Brown Effect

During the 1920's, Thomas Townsend Brown was experimenting with an X-ray tube known as a "Coolidge tube", which was invented in 1913 by the American physical chemist William D. Coolidge [1]. Brown found that the Coolidge tube exhibited a net force (a thrust) when it was turned on. He believed that he had discovered a new principle of electromagnetism and gravity. Brown applied for a British patent on April 15, 1927, which was issued on November 15, 1928 as Patent No. 300,311, entitled, "Method of Producing Force or Motion." The patent and its figures clearly describe Brown's early work on forces on asymmetric capacitors, although the electromagnetic concepts are mixed with gravitational concepts.[37]

The discovery of the Biefeld-Brown effect is generally credited to Thomas Townsend Brown. However, it is also named in honor of Brown's mentor, Dr. Paul Alfred Biefeld, a professor of physics and astronomy at Denison University in Granville, Ohio, where Brown was a lab assistant in electronics in the Department of Physics. During the 1920's, Biefeld and Brown together experimented on capacitors.

In order to find a technical description of the Biefeld-Brown effect, we performed a search of the standard article literature, and found no

---

[37] It states, "This invention relates to a method of **controlling gravitation** and for deriving power therefrom, and to a method for producing linear force or motion. The method is fundamentally electrical." The complete patent is in the first volume of this series. –Ed note

references to this effect. It is prudent to ask whether this effect is real or rumor. On the other hand, the Internet is full of discussions and references to this effect, including citations of patents issued [1] (see also Appendix A). In fact, patents seem to be the only official publications that describe this effect.

On July 3,1957, Brown filed another patent entitled "Electrokinetic Apparatus", and was issued a US Patent No. 2,949,550 on August 16, 1960. The effect in this patent is described more lucidly than his previous patent No. 300,311, of November 15, 1928. In this 1960 patent, entitled "Electrokinetic Apparatus," Brown makes no reference to gravitational effects:

The invention utilizes a heretofore unknown electro-kinetic phenomenon which I have discovered; namely, that when a pair of electrodes of appropriate form are held in a certain fixed spaced relation to each other and immersed in a dielectric medium and then oppositely charged to an appropriate degree, a force is produced tending to move the pair of electrodes through the medium. The invention is concerned primarily with certain apparatus for utilizing such phenomenon in various manners to be described.

Priorly, intervening electrokinetic apparatus has been employed to convert electrical energy to mechanical energy and then to convert the mechanical energy to the required force. Except for the insignificantly small forces of electrostatic attraction and repulsion, electrical energy has not been used for the direct production of force and motion.

Fig. 1 Excerpt from Thomas Townsend Brown US Patent No. 2,949,550 entitled "Electrokinetic Apparatus", issued on August 16, 1960.

The claims, as well as the drawings in this patent clearly show that Brown had conceived that the force developed on an asymmetrical capacitor could be used for vehicle propulsion. His drawings in this patent are strikingly similar to some of the capacitors designs on the Internet today. In this 1960 patent, entitled "Electrokinetic Apparatus," Brown gives the clearest explanation of the physics of the Biefeld-Brown effect. Brown makes several important statements, including:

- the greatest force on the capacitor is created when the small electrode is positive
- the effect occurs in a dielectric medium (air)
- the effect can be used for vehicle propulsion, or as a pump of dielectric fluid
- Brown's understanding of the effect, in terms of ionic motion the detailed physics of the effect is not understood

Next, we reproduce Brown's first two figures and partial text explaining the effect:

3

I have discovered that when apparatus of the character just described is immersed in a dielectric medium, as for example, the ordinary air of the atmosphere, there is produced a force tending to move the entire assembly through the medium, and this force is applied in such a direction as to tend to move the body 20 toward the leading electrode 21. This force produces relative motion between the apparatus and the surrounding fluid dielectric. Thus, if the apparatus is held in a fixed position, the dielectric medium is caused to move past the apparatus and to this extent the apparatus may be considered as analogous to a pump or fan. Conversely, if the apparatus is free to move, the relative motion between the medium and the apparatus results in a forward motion of the apparatus, and it is thus seen that the apparatus is a self-propulsive device.

While the phenomenon just described has been observed and its existence confirmed by repeated experiment, the principles involved are not completely understood. It has been determined that the greatest forces are developed when the leading electrode is made positive with respect to the body 20, and it is accordingly thought that in the immediate vicinity of the electrode 21 where the potential gradient is very high, free electrons are stripped off of the atoms and molecules of the surrounding medium. These electrons migrate to the positive electrode 21 where they are collected. This removal of free electrons leaves the respective atoms and molecules positively charged and such charged atoms and molecules are accordingly repelled from the positive electrode 21 and attracted toward the negative electrode 20. The paths of movement of these positively charged particles appear to be of the nature represented by the lines 27 in Figure 2.

It appears that upon reaching or closely approaching the surface of the body 20, the positively charged atoms and molecules have their positive charges neutralized by the capture of electrons from the body 20 and in many cases, it may be that excess electrons are captured whereby to give such atoms and molecules a negative charge so that they are actually repelled from the body 20.

It will be appreciated that the mass of each of the individual electrons is approximately one two-thousandths the mass of the hydrogen atom and is accordingly negligible as compared with the mass of the atoms and molecules of the medium from which they are taken. The principal forces involved therefore are the forces involved in moving the charged atoms and molecules from the region of the positive electrode 21 to and beyond the negatively charged body 20. The force so exerted by the system on those atoms and molecules not only produces a flow of the medium relative to the apparatus, but, of course, results in a like force on the system tending to move the entire system in the opposite direction; that is, to the left as viewed in Figure 1 of the drawing.

The above suggested explanation of the mode of operation of the device is supported by observation of the fact that the dimensions and potentials utilized must be adjusted to produce the required electric field and the resulting propulsive force. Actually I have found that the potential gradient must be below that value required to produce a visible corona since corona is objectionable inasmuch as it represents losses through the medium. My experiments have indicated that the electrode 21 may be of small diameter for the lower voltage ranges, i.e. below 125 kv. while above this voltage, rod or hollow pipe electrodes are preferred. These large electrodes are preferred for the higher voltages since sharp points or edges are eliminated which at these elevated potentials would produce losses thus diminishing the thrust. For example, electrodes to be operated at potentials below 125 kv. may be made from small gauge wire only large enough to provide the required mechanical rigidity while

4

electrodes to be operated at potentials above 125 kv. may be hollow pipes or rods having a diameter of ¼ to ½ inch.

In Figure 3, I have illustrated the manner in which a plurality of assemblies, such as are shown in Figure 1, may be interconnected for joint operation. As may be seen from Figure 3, a plurality of such assemblies are placed in spaced side-by-side relation. They may be held fixed in such spaced relation through the use of a plurality of tie rods 28 and interposed spacers (not shown) placed between adjacent plates 20. The assembly of plates 20 may be electrically interconnected by a bus bar or similar conductor 29 to which the negative lead 25 is connected. In a similar way, the plurality of positive leading electrodes 21 may be held in appropriately spaced relation to each other by fastening their ends to pairs of bus bars 30 and 31, to the latter of which the positive lead 26 is connected. The assembly of leading electrodes 21 may be held in spaced relation to the assembly of body members 20 by an appropriate arrangement of the supports 22.

In Figure 4, I have illustrated diagrammatically an arrangement of parts for producing a reversible action; that is, permitting the direction of the propulsive force to be reversed. The apparatus is similar to that shown in Figure 1, differing therefrom in utilizing a pair of leading electrodes 21f and 21r spaced by means of spacers 22 from the front and rear edges 23f and 23r of the body member 20 in a manner similar to that described with reference to the supports 22 in Figure 1. The source 24 of high voltage electrical potential has its negative terminal connected to the body 20 as by means of the aforementioned conductor 25. The positive terminal is connected as by means of the conductor 26 to the blade 27 of a single-pole, double-throw switch, serving in one position to connect the conductor 26 to a conductor 26f which is in turn connected to the forward electrode 21f and arranged in its opposite position to connect the conductor 26 to a conductor 26r which is in turn connected to the reverse electrode 21r.

It will be seen that with the switch 27 in the position shown in Figure 4, the apparatus will operate in the manner described in connection with Figure 1, causing the assembly to move to the left as viewed in Figure 4. By throwing the switch 27 to the opposite position, the direction of the forces produced are reversed and the device moves to the right as viewed in Figure 4.

In Figure 5, I have illustrated the principles of the invention as embodied in a simple form of mobile vehicle. This device includes a body member 50 which is preferably of the form of a circular disc somewhat thicker in its center than at its edges. The disc 50 constitutes one of the electrodes and is the equivalent of the body member 20 referred to in connection with Figure 1. A leading electrode 51 in the form of a wire or similar small diameter conductor is supported from the body 50 by a plurality of insulating supports 52 in uniform spaced parallel relation to a leading edge portion 53 of the body 50. A skirt or similar fairing 54 may be carried by the body 50 to round out the entire structure so as to provide a device which is substantially circular in plan. A source of high voltage electrical potential 55 is provided with its negative terminal connected as indicated at 56 to the body 50 and its positive terminal connected as indicated at 57 to the leading electrode 51.

The device operates in the same manner as the apparatus shown in Figure 1 to produce a force tending to move the entire assembly through the surrounding medium to the left as viewed in Figure 5 of the drawing.

Referring now to Figure 6, there is depicted an illustrative embodiment of this invention in which a pair of mobile vehicles, such as depicted in Figure 5, are shown suspended from the terminals of arm 40, which arm is supported at its midpoint by a vertical column 41. High voltage source 55 is shown connected through wires

Fig. 2 Excerpt from Thomas Townsend Brown US Patent No. 2949550 entitled "Electrokinetic Apparatus", issued on August 16, 1960.

31

FIG. 1   *FIG. 2*

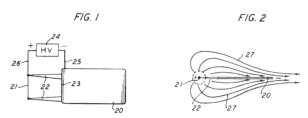

Fig. 3 Figure excerpt from Thomas Townsend Brown US Patent No. 2949550 entitled "Electrokinetic Apparatus", issued on August 16, 1960.

Soon after Brown's 1957 filing for the above patent, on May 12, 1958, A.H. Bahnson Jr. filed for an improved patent entitled "Electrical thrust producing device," which was granted a US Patent No. 2,958,790 on November 1, 1960.

On July 3, 1957, Brown filed another patent (granted on Jan 23, 1962, as US patent No. 3,018,394) for an "Electrokinetic Transducer." This patent deals with the inverse effect: when a dielectric medium is made to move between high voltage electrodes, there is a change in the voltage on the electrodes. (This is reminiscent of Faraday's law of induction.) Quoting from the 1962 patent by Thomas Townsend Brown:

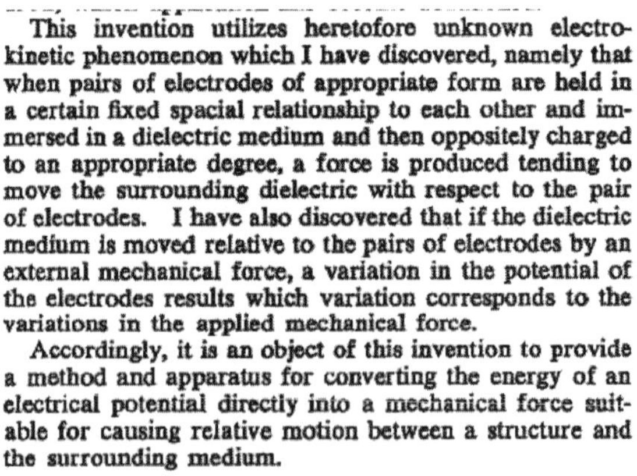

This invention utilizes heretofore unknown electrokinetic phenomenon which I have discovered, namely that when pairs of electrodes of appropriate form are held in a certain fixed spacial relationship to each other and immersed in a dielectric medium and then oppositely charged to an appropriate degree, a force is produced tending to move the surrounding dielectric with respect to the pair of electrodes. I have also discovered that if the dielectric medium is moved relative to the pairs of electrodes by an external mechanical force, a variation in the potential of the electrodes results which variation corresponds to the variations in the applied mechanical force.

Accordingly, it is an object of this invention to provide a method and apparatus for converting the energy of an electrical potential directly into a mechanical force suitable for causing relative motion between a structure and the surrounding medium.

Fig. 4

Excerpt from Thomas Townsend Brown US patent No. 3018394 entitled "Electrokinetic Transducer," issued on January 23, 1962.

32

Until this time, the net force on an asymmetric capacitor was reported as occurring when the capacitor was in a dielectric medium. On May 9, 1958, Brown filed for another patent (improving upon his previous work) entitled "Electrokinetic Apparatus." The patent was issued on June 1, 1965 as Patent No. 3,187,206. The significance of this new patent is that it describes the existence of a net force on the asymmetric capacitor as <u>occurring even in vacuum</u>. Brown states that, "The propelling force however is not reduced to zero when all environmental bodies are removed beyond the apparent effective range of the electric field." Here is a quote from the patent:

3,187,206
ELECTROKINETIC APPARATUS
Thomas Townsend Brown, Walkertown, N.C., assignor, by mesne assignments, to Electrokinetics, Inc., a corporation of Pennsylvania
Filed May 9, 1958, Ser. No. 734,342
23 Claims. (Cl. 310—5)

This invention relates to an electrical device for producing thrust by the direct operation of electrical fields.

I have discovered that a shaped electrical field may be employed to propel a device relative to its surroundings in a manner which is both novel and useful. Mechanical forces are created which move the device continuously in one direction while the masses making up the environment move in the opposite direction.

When the device is operated in a dielectric fluid medium, such as air, the forces of reaction appear to be present in that medium as well as on all solid material bodies making up the physical environment.

In a vacuum, the reaction forces appear on the solid environmental bodies, such as the walls of the vacuum chamber. The propelling force however is not reduced to zero when all environmental bodies are removed beyond the apparent effective range of the electrical field.

By attaching a pair of electrodes to opposite ends of a dielectric member and connecting a source of high electrostatic potential to these electrodes, a force is produced in the direction of one electrode provided that electrode is of such configuration to cause the lines-of-force to converge steeply upon the other electrode. The force, therefore, is in a direction from the region of high flux density toward the region of low flux density, generally in the direction through the axis of the electrodes. The thrust produced by such a device is present if the electrostatic field gradient between the two electrodes is non-linear. This non-linearity of gradient may result from a difference in the configuration of the electrodes, from the electrical potential and/or polarity of adjacent bodies, from the shape of the dielectric member, from a gradient in the density, electric conductivity, electric permittivity and magnetic permeability of the dielectric member or a combination of these factors.

Fig. 5
Excerpt from Thomas Townsend Brown Patent No. 3,187,206, entitled, "Electrokinetic Apparatus," issued on June 1, 1965.

33

In the above patent, Brown reports that the asymmetric capacitor does show a net force, even in vacuum. However, a present, there is little experimental evidence, except for two reports [2], which do not explain the origin of the observed force. If the Biefeld-Brown effect is to be understood on a firm basis, it is imperative to determine whether the effect occurs in vacuum. Enclosed in Appendix B, is my email correspondence with J. Naudin, where Naudin quotes from a letter by Thomas Townsend Brown, who discusses the effect in vacuum. The main question to be answered is: what is the physical mechanism that is responsible for the net force on an asymmetric capacitor? The answer to this question may depend on whether the asymmetric capacitor is in a polarizable medium, in air, or in vacuum. However, to date the physical mechanism is unknown, and until it is understood, it will be impossible to determine its potential for practical applications.

3. Preliminary Experiments at Army Research Laboratory

The Biefeld-Brown effect is reported many places on the Internet, however, it is not described in any physics journals. Therefore, we decided to verify that the effect was real. C. Fazi (Army Research Laboratory (ARL)) and T. Bahder (ARL) have fabricated three simple asymmetric capacitors, using the designs reported on the Internet [1]. In all three cases, we have verified that a net force is exerted on the capacitors when a high DC voltage is applied to the electrodes.

The three asymmetric capacitors that we tested had different geometries, but they all had the common feature that one electrode was thin and the other very wide (asymmetric dimensions). Also, a suspended wire, representing a capacitor with the second electrode at infinity, showed lift.

Our first model was made by Tom Bahder, and was triangular shape, which is a typical construction reported on the Internet (see Figure 6). One electrode is made from thin 38 gauge (0.005 mil) wire, and the other electrode is made from ordinary Aluminum foil. The capacitor is approximately 20 cm on a side, the foil sides are 20 cm × 4 cm, and them distance of the top of the foil to the thin wire electrode is 3 cm. The foil and wire are supported by a Balsa wood frame, so that the whole capacitor is very light, approximately 5 grams. Initially, we made the Balsa wood frame too heavy (capacitor weight about 7 grams), and later we cut away much of the frame to lighten the construction to about 5 grams. We found that in order to demonstrate the lifting effect, the capacitor must be made of minimum weight. (Typical weights reported on the Internet for the design in Figure 6 are 2.3 grams to 4 grams.)

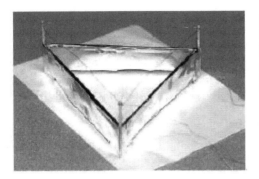

Fig. 6
Our first attempt at making an asymmetric capacitor (a "lifter"), according to the specifications given by J. Naudin on Internet web site http://jnaudin.free.fr/.

When about 37 kV was applied to the capacitor in Figure 6, the current was about 1.5 mA. The capacitor lifted off its resting surface. However, this capacitor was not a vigorous flier, as reported by others on the Internet. One problem that occurred was arcing from the thin wire electrode to the foil. The thin wire electrode was too close to the foil. We have found that arcing reduces the force developed on the capacitor. Also, compared to other constructions, ours was too heavy, 5 grams. We found that a ground plane beneath the capacitor is not essential for the lifting force to exceed the capacitor's weight.

Consequently, we decided to make a second version of an asymmetric capacitor, using a styrofoam lunch box and plastic drinking straws from the ARL cafeteria (See Figure 7). The capacitor had a square geometry 18 cm × 20 cm. The distance of the thin wire (38 gauge) to the foil was adjustable, and we found that making a 6 cm gap resulted in little arcing. When 30 kV was applied, the capacitor drew about 1.5 mA, and hovered vigorously above the floor.

A question occurred: is the toroidal (closed circular) geometry of the capacitor electrodes essential to the lifting effect that we have observed. Consequently, Tom Bahder made a flat-shaped, or wing-shaped, capacitor as shown in Figure 8. This capacitor was made from two (red) plastic coffee stirrers and a (clear) plastic drinking straw to support the Aluminum foil. The significance of the clear plastic straw was that the foil could be wrapped over it, thereby avoiding sharp foil edges that would lead to corona discharge or arcing. The dimensions of the foil on this capacitor were 20 cm × 4 cm, as shown in Figure 8. The distance between the thin wire electrode (38 gauge wire) and edge of the foil was 6.3 cm. This capacitor showed a net force on it when about 30 kV was applied, drawing about 500 mA. The force on this capacitor greatly exceeded its weight, so much so that it would vigorously fly into the air when the voltage was increased from zero. Therefore, we have concluded that the closed geometry of the electrodes is not a factor in the net force on an asymmetric capacitor. Furthermore, the force on the capacitor always appeared in the direction toward the small electrode – independent of the orientation of the capacitor with respect to the plane of the Earth's surface. The significance of this observation is that the force has nothing to do with the gravitational field of the Earth, and nothing to do with the electric potential of the Earth's atmosphere.

Figure 7: The second attempt at making a lighter asymmetric capacitor.

There are numerous claims on the Internet that asymmetric capacitors are anti-gravity devices, or devices that demonstrate that there is an interaction of gravity with electric phenomena, called. The thin wire electrode must be at a sufficient distance away from the foil so that arcing does not occur from the thin wire electrode to the foil, at the operating voltage. In fact, in our first model, shown in Figure 6, the 3 cm gap from to of the foil to thin wire electrode was not sufficiently large, and significant arcing occurred. We have found that when arcing occurs, there is little net force on the capacitor. An essential part of the design of the capacitor is that the edges of the foil, nearest to the thin wire, must be rounded (over the supporting Balsa wood, or plastic straw, frame) to prevent arcing or corona discharge at sharp foil edges (which are closest to the thin wire). The capacitor in Figure 6 showed improved lift when rounded foil was put over the foil electrode closest to the thin wire, thereby smoothing-over the sharp

36

foil edges. Physically, this means that the radius of curvature of the foil nearest to the small wire electrode was made larger, creating a greater asymmetry in radii of curvature of the two electrodes. When operated in air, the asymmetric capacitors exhibit a net force toward the smaller conductor, and in all three capacitors, we found that this force is independent of the DC voltage polarity. The detailed shape of the capacitor seem immaterial, as long as there is a large asymmetry between the characteristic size of the two electrodes. A suspended thin wire (approximately 12 in length) also showed lift with about 37 kV and 1 mA current (see Figure 8).

When the asymmetric capacitors have an applied DC voltage, and they are producing a net force in air, they all emit a peculiar hissing sound with pitch varying with the applied voltage. This sound is similar to static on a television or radio set when it is not tuned to a good channel. We believe that this sound may be a clue to the mechanism responsible for the net force.

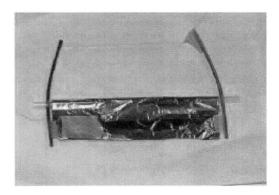

Fig. 8

Flat shaped (or wing-shaped) asymmetric capacitor used to test whether closed electrode geometry is needed.

The simplest capacitor configuration consists of a suspended thin wire from the hot electrode of the high voltage power supply (see Figure 9).

To observe the wire movement, a small piece of transparent tape was attached at the lower end of the thin wire. From a vertical position, the wire lifted, as shown in Figure 10 by as much as 30 degrees, once the high voltage approached 35kV. The usual air breakdown hissing sound of the other capacitors was heard when current drain reached about 1mA. Actually the wire did not remain suspended, but oscillated back and forth approximately 60 degrees from vertical, and the hissing pitch followed the oscillation period with amplitude and frequency changes.

Fig. 9 The capacitor consisting of a single wire. No bias applied.

Fig. 10 The wire capacitor showing displacement from the vertical. 35 kV applied.

Without the piece of tape at the end, the wire did not lift as much and the sound was considerably weaker. The piece of tape seems to increase the capacitance and or the air ionization. This suspended wire configuration can be viewed also as a capacitor surrounded by the ground system located several feet away (metallic benches, floor and ceiling). As in the other capacitor experiments, it also did not exhibit a polarity dependence.

4.   Previously Proposed Explanations for the Biefeld-Brown Force

There are two proposed explanations for the Biefeld-Brown force. Both of these have been discussed on the Internet in various places. The first proposed scheme is that there exists an ionic wind in the high field region between the capacitor electrodes, and that this ionic wind causes the electrodes to move as a result of the momentum recoil. This scheme, described in **Section A** below, leads to a force that is incorrect by at least three orders order of magnitude compared to what is observed. (This scheme also assumes ballistic transport of charges in the atmosphere between electrodes of the capacitor, and it is is known that instead drift current exists.) In **Section B** below, we present the

38

second scheme, which assumes that a drift current exists between the capacitor plates.

This scheme is basically a scaling argument, and not a detailed treatment of the force. In this scheme, the order of magnitude of the force on an asymmetric capacitor is correct. However, this scheme is only a scaling theory. Finally, in Section 5 below, we present our thermodynamic treatment of the force on an asymmetric capacitor.

A. Ionic Wind, Force Too Small

The most common explanation for the net force on an asymmetric capacitor invokes ionic wind. Under a high voltage DC bias, ions are thought to be accelerated by the high potential difference between electrodes, and the recoil force is observed on an asymmetric capacitor. A simple upper limit on the ion wind force shows that the ion wind effect is a factor of at least three orders of magnitude too small. Consider a capacitor that operates at voltage V. Charged particles of mass m, having charge q, such as electrons or (heavy) ions, are accelerated to a velocity v, having a kinetic energy

$$\frac{1}{2}mv^2 = qV \qquad (1)$$

The force exerted on an asymmetric capacitor is given by the rate of change of momentum

$$F = mv\frac{I}{q} \qquad (2)$$

where I is the current flowing through the capacitor gap, and we assume that all the ionic momentum mv, is transferred to the capacitor when the charged particles leave an electrode. Also, we assume that none of this momentum is captured at the other electrode. This is a gross over-estimation of the force due to ionic effects, so Eq. (2) is an upper limit to the ionic force.

Solving Eq.(1) for the velocity, and using it in Eq.(2) gives the upper limit on the force due to ionic wind

$$F = I\sqrt{\frac{2mV}{q}} \qquad (3)$$

When the force F is equal to the weight of an object Mg, where g is the acceleration due to gravity, the force will lift a mass

39

$$M = \frac{I}{g}\sqrt{\frac{2mV}{q}} \tag{4}$$

If we assume that electrons are the charged particles responsible for force of the ionic wind, then we must use mass $m = 9.1 \times 10^{-31}$ kg. Substituting typical experimental numbers into Eq. (4), we find that the ionic wind can lift a mass

$$M = \frac{10^{-3}\,A}{10\,\frac{m}{s^2}}\sqrt{\frac{2(9.1\times10^{-31}\,kg)(40kV)}{1.6\times10^{-19}\,C}} = 6.8\times10^{-5}\,grams \tag{5}$$

The typical weight of an asymmetric capacitor is on the order of 5 grams, so this force is too small by 5 orders of magnitude.

Another possibility is that heavy ions (from the air or stripped off the wire) are responsible for the ionic wind. As the heaviest ions around, assume that Cu is being stripped from the wire. Using Cu for the ions, the mass of the ions is 63.55, which is the atomic mass of Cu, and m_p is the mass of a proton. The weight that could be lifted with Cu ionic wind is then (upper limit):

$$M = \frac{10^{-3}\,A}{10\,\frac{m}{s^2}}\sqrt{\frac{(2)(63.55)(1.67\times10^{-27}\,kg)(40kV)}{1.6\times10-19C}} = 2\times10^{-3}\,gm \tag{6}$$

Again, this value is three orders of magnitude too small to account for lifting a capacitor with a mass of 3 to 5 grams. Therefore, the ionic wind contribution is too small, by at least three orders of magnitude, to account for the observed force on an asymmetric capacitor. While the force of the ionic wind computed above is too small to explain the experiments in air, it should be noted that this effect will operate in vacuum, and may contribute to the overall force on a capacitor.

B. The Ion Drift Picture: Scaling Theory of Force

In the previous section, we computed an upper limit to the force on a capacitor due to ionic wind effects. Ionic wind is a ballistic flow of charges from one electrode to the other. Clearly the force due to ionic wind is at least three orders of magnitude too small to account for the observed force on an asymmetric capacitor (in air). There is another type of classical transport: drift of charge carriers in an electric field. In the case of drift, the carriers do not have ballistic trajectories, instead they experience collisions on their paths between electrodes. However, due to the presence of an electric field, the carriers have a net motion toward the opposite electrode. This type of transport picture is more accurate (than ballistic ionic wind) for a capacitor whose gap contains air. Drift transport is used by Evgenij Barsoukov to explain the net force on an asymmetric capacitor [3]

The general picture of the physics is that the positive and negative electrodes of the capacitor are charged and that these charges experience different forces, because the electric field surrounding the capacitor is non-uniform (See Figure 10).

The electric field surrounding the capacitor is created by the potential applied to the capacitor electrodes and partial ionization of air into positive ions and electrons. These charge carriers experience drift and diffusion in the resulting electric field. The battery supplies the energy that is dissipated by transport of carriers in the electric field. The electric field is particularly complicated because it is the result of a steady state: the interplay between the dynamics of ionization of the air in the high-field region surrounding the electrodes, and charge transport (drift and diffusion of positive and negative carriers) in the resulting electric field.

If the capacitor is surrounded by vacuum (rather than a dielectric, such as air), the net force F on the asymmetric capacitor can be computed by the sum of two surface integrals, one over the surface of the positive electrode and one over the surface of the negative electrode [4]:

$$F = \frac{1}{2}\varepsilon_o \left[\oint_{S_+} E^2 n \, dS + \oint_{S_-} E^2 n \, dS \right] \qquad (7)$$

where E is the electric field due to charges in the ionized air between electrodes (excluding the field due to surface charge on the capacitor electrodes) and S_+ and $S-$ are the positive and negative electrode surfaces of the capacitor. As stated above, the complexity of the calculation is contained in computing the electric field E. In Section 5, we give an expression for the net force on the capacitor assuming that it is surrounded by a dielectric, such as air.

An alternative but equivalent picture is that the capacitor is an electric dipole in an non-uniform electric field that it has produced, and the ions form a molasses, due to their high mass and resulting low mobility. We will develop both pictures below in scaling arguments.

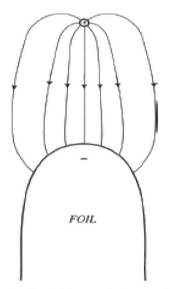

Fig. 11
Schematic diagram of the side view of electric field for the asymmetric capacitor in Figure 8.

The electric field around the small wire electrode is much stronger than the field around the foil (see Figure 8 and 11). In our experiments, there is a big difference in the radii of curvature of the two capacitor electrodes: the thin wire electrode has a radius $r_1 = .0025$ inch, and the edge of the foil has a radius of curvature of $r_2 = .125$ inch. This difference in curvature leads to an electric field with a strong gradient. The ratios of electric fields at the thin wire electrode to that at the rounded edge of the foil is inversely proportional to the square of the radii of curvatures: $E_1/E_2 = (r_1/r_2)^2 \approx 2500$. However, the applied voltage is on the order of 30 kV, over a gap of 6 cm, so an electric field of magnitude 2500 x 30 kV / 6 cm $\approx 10^7$ V/cm would not be supported in air. It is clear that screening of the electric field is occurring due to the dielectric effects of charged air ions and electrons, as well as polarized air atoms. When a positive high voltage is applied to the asymmetric capacitor, ionization of air atoms, such as Nitrogen, probably occurs first near the thin wire electrode. The ionization of Nitrogen atoms leads to free electrons and ions near the small electrode. The electron mobility is significantly larger for electrons than for Nitrogen ions. This can be expected since the current density

$J = \sigma E = n e v$ where $\sigma = n e^2 \tau / m$ is the electrical conductivity, n is charge density, t is the scattering time, and the mean drift velocity $v = m E$. So the mobility behaves as $\mu = e \tau / m$. Since electrons are three orders of magnitude more massive than ions, it is expected that they are correspondingly more mobile. Experimentally, it is found that the electron mobility in air at atmospheric pressure and electric field $E = 10^4$ Volt/cm is approximately [5]

$$\mu_e = 620 \frac{cm^2}{Volt \cdot sec} \tag{8}$$

The mobility of N_2 ions in air is [6]

$$\mu_{N_2} = 2.5 \frac{cm^2}{Volt \cdot sec} \tag{9}$$

Therefore, the physical picture is that in the high field region the electrons, with their high mobility, are swept out by the electric field, toward the thin wire electrode and cause dynamic screening of the potential. (Dielectric screening due to polarized air atoms will also take place.) However, the massive (probably positive) ions are less mobile and are left behind in a plasma surrounding the thin wire electrode. A scaling argument can be made as follows. The lower foil conductor feels a force F of magnitude

$$F = Q \frac{V}{l} \tag{10}$$

where Q is the charge on the foil electrode V is the voltage between the capacitor conductors, and l is the length of the gap between thin wire electrode and foil. The charge Q and voltage V are quantities that are actually present with shielding taking place. The negative charge on the foil, $-Q$, can be approximated in terms of the measured current, I ~1mA, by saying that all the carriers are swept out in a time t

$$I = \frac{Q}{t} = Q \frac{v}{l} \tag{11}$$

43

where t is the time for carriers to move across the capacitor gap l, if they are traveling at an average drift velocity v. Note that the measured current is due to the electrons. Eliminating the charge Q from Eq. (10) and (11) leads to an expression for the net force on the capacitor

$$F = I \frac{V}{v} \tag{12}$$

In Eq. (11), the current I is a measured quantity, the voltage V is on the order of 30 kV, and the drift velocity for electrons is [5]

$$v_e = 6.2 \times 10^6 \frac{cm}{sec} \tag{13}$$

Alternatively, the electron drift velocity, v_e, can be expressed in terms of the mobility, m_e given in Eq. (8), and electric field, E. The net force on the asymmetric capacitor is then given by

$$F = I \frac{V}{\mu E} = I \frac{l}{\mu} \tag{14}$$

where we again used $E = V / l$. Using the value of electron mobility in Eq. (8), the net force becomes

$$F = I \frac{l}{\mu} = \frac{\left(10^{-3}\,A\right)\left(0.04m\right)}{\left(620 \dfrac{cm^2}{Volt \cdot sec}\right)\left(10^{-2} \dfrac{m}{cm}\right)^2} = 6.4 \times 10^{-4}\,N \tag{15}$$

The force in Eq.(14), due to the drift of electrons, could lift a mass M

$$M = \frac{F}{g} = \frac{6.4 \times 10^{-4}\,N}{9.8 \dfrac{m}{s^2}} = 0.065\,gram \tag{16}$$

The typical asymmetric capacitor has a mass that is <u>two orders of magnitude greater</u>. *Consequently, drift of electrons cannot explain the observed force on the capacitor.*

An alternative to using the value of electron mobility is to use the smaller value of ionic mobility. (This will lead to a larger force

44

because the force in Eq. (14) is inversely proportional to the mobility.) It is not clear how this can be justified, however, the numbers come out closer to what is observed. Using the mobility of nitrogen ions in air, given in Eq.(8), the force becomes

$$F = I \frac{l}{\mu} = \frac{\left(10^{-3} A\right)\left(0.04m\right)}{\left(2.5 \frac{cm^2}{Volt \cdot \sec}\right)\left(10^{-2} \frac{m}{cm}\right)^2} = 0.16N \tag{17}$$

The force in Eq. (16), due to the drift of Nitrogen ions, could lift a mass M

$$M = \frac{F}{g} = \frac{0.16N}{9.8 \frac{m}{s^2}} = 16 \ grams \tag{18}$$

The force on the capacitor, given in Eq. (18), is within a factor of 3, assuming a capacitor of mass 5 grams. However, it is difficult to see why ion mobility is the appropriate quantity to use in the derivation of the force.[38]

As an alternative derivation of the scaling Eq. (14), consider the asymmetric capacitor as being essentially an electric dipole of magnitude,

$$\left| \mathbf{p} \right| = p = Ql \tag{19}$$

where Q is the charge on one plate and l is the average effective separation between plates. When a high voltage is applied to the asymmetric capacitor (assume positive voltage on the thin wire and negative on the foil), the high electric field around the thin wire ionizes the atoms of the air. There is comparatively little ionization near the foil due to the lower magnitude electric field near the foil. The ionized atoms around the foil form a plasma, consisting of charged electrons and positively charged ions. The force on the capacitor must scale like

$$\mathbf{F} = \nabla\left(\mathbf{p} \cdot \mathbf{E}\right) \tag{20}$$

---

[38] This is especially true if the Biefeld-Brown effect functions in a vacuum, where the density of ions is extremely low (see Appendix B) – Ed note

45

where $\mathbf{E}$ is the electric field. The gradient operates on the electric field, producing a magnitude $dE/dx \sim E / l$. Using this value in Eq.(18), together with the size of the dipole, leads to a force on the capacitor

$$F = Q\frac{V}{l} \approx \frac{I\,l}{v}\cdot\frac{V}{l} = I\frac{V}{v} \tag{21}$$

which is identical to Eq. (14).

From the scaling derivations presented above, it is clear that electron drift current leads to a force on the capacitor that is too small. Using the value of mobility appropriate for (nitrogen) ions leads to a force magnitude in agreement with experiment. However, it is not clear why the mobility of the ions should be used in the calculation.

Note that the force, given by Eq. (14), *scales inversely with the mobility m.* If the ions are responsible for providing the required small mobility, then the picture is that the ions are like a low-mobility molasses, which provides a large space charge to attract the negatively charged foil electrode. As soon as the foil electrode moves toward the positive ion cloud, another positive ionic cloud is set up around the thin electrode, using the energy from the voltage source. In this way, the dipole (asymmetric capacitor) moves in the non-uniform electric field that it has created. *Physically, this is a compelling picture.* However, much work must be done (experimentally and theoretically) to fill in important details to determine if this picture has any merit.

5. Thermodynamic Analysis of the Biefeld-Brown Force

In this section, we present our hypothesis that the Biefeld-Brown force, generated on an asymmetric capacitor, can be described by the thermodynamics of a fluid dielectric in an external electric field produced by charged conductors. The (partially ionized) air between capacitor electrodes is the fluid dielectric. Although the air is partially ionized, we assume that this fluid dielectric is close to neutral on the macroscopic scale. The charged conductors are the asymmetric electrodes of the capacitor. The battery provides the charge on the electrodes and the energy sustain the electric field in the air (dielectric) surrounding the capacitor electrodes.

The total system is composed of three parts: the partially ionized air dielectric, the metal electrodes of the capacitor and the battery (voltage source), and the electromagnetic field. The battery is simply a large reservoir of charge. The total momentum (including the electromagnetic field) of this system must be constant [9]

$$\mathbf{P}_{dielectric} + \mathbf{P}_{electrodes} + \mathbf{P}_{field} = constant \tag{22}$$

where $\mathbf{P}_{dielectric}$ is the momentum of the fluid dielectric (air in the capacitor gap and surrounding region), $\mathbf{P}_{electrodes}$ is the momentum of the metallic electrodes and battery, and $\mathbf{P}_{field}$ is the momentum of the electromagnetic field. Taking the time derivative of Eq. (22), the forces must sum to zero

$$\mathbf{F}_{dielectric} + \mathbf{F}_{electrodes} + \frac{d\,\mathbf{P}_{field}}{dt} = 0 \tag{23}$$

As far as the electric field is concerned, its total momentum changes little during the operation of the capacitor, because the field is in a steady state; energy is supplied by the battery (charge reservoir). So we set the rate of change of field momentum to zero, giving a relation between the force on the electrodes and the dielectric:

$$\mathbf{F}_{electrodes} = -\mathbf{F}_{dielectric} \tag{24}$$

A lengthy derivation based on thermodynamic arguments leads to an expression for the stress tensor, σ_{ik} , for a dielectric medium in an electric field [4,7,8]

$$\sigma_{ik} = \left[\widetilde{F} - \rho \left(\frac{\partial \widetilde{F}}{\partial \rho} \right)_{T,\mathbf{E}} \right] \delta_{ik} + E_i D_k \tag{25}$$

where the free energy is a function of the fluid density, r, temperature, T, and electric field $\mathbf{E}$. The differential of the free energy is given by

$$d\,\widetilde{F} = -S\,dT + \zeta\,d\rho - \mathbf{D} \cdot d\mathbf{E} \tag{26}$$

where S is the entropy, $\mathbf{D}$ is the electric induction vector, and ζ is the chemical potential per unit mass [4]. Equation (25) is valid for any constitutive relation between $\mathbf{D}$ and $\mathbf{E}$. We assume that the air in between the capacitor plates is an isotropic, but nonlinear, polarizable medium, due to the high electric fields between plates. Therefore, we take the relation between **D** and **E** to be

$$\mathbf{D} = \varepsilon(E)\,\mathbf{E} \tag{27}$$

where $\varepsilon(E)$ is a scalar dielectric function that depends on the magnitude of the electric field, $E = |\mathbf{E}|$, the temperature, T, and the density of the fluid, r. We have suppressed the dependence of ε on T and ρ for brevity. The dielectric function $\varepsilon(E)$ depends on position through the variables T and ρ and because the medium (air) between capacitor plates is assumed to be non-uniform. Inserting Eq. (27) into Eq. (26), we integrate the free energy along a path from $E = 0$ to some finite value of E obtaining

$$\widetilde{F}(\rho,T,\mathbf{E}) = \widetilde{F}_o(\rho,T) - \frac{1}{2}\varepsilon_{\text{eff}}E^2 \tag{28}$$

where ε_{eff} is an effective (averaged) dielectric constant given by

$$\varepsilon_{\text{eff}} = \frac{1}{E^2}\int_0^{E^2}\varepsilon\!\left(\sqrt{\xi}\right)d\xi \tag{29}$$

where ξ is a dummy integration variable. The dielectric constant ε_{eff} depends on spatial position (because of ε), on T, ρ, and on electric field magnitude E.

The body force per unit volume on the dielectric f_i, is given by the divergence of the stress tensor

$$f_i = \frac{\partial \sigma_{ik}}{\partial x_k} \tag{30}$$

where there is an implied sum over the repeated index k. Performing the indicated differentiations, we obtain an expression for the for body force [4]

$$f = -\nabla P_o(\rho,T) + \nabla\!\left[\frac{E^2\rho}{2}\!\left(\frac{\partial\varepsilon_{\text{eff}}}{\partial\rho}\right)_{T,E}\right] - \frac{E^2\nabla\varepsilon_{\text{eff}}}{2} + \frac{(\varepsilon - \varepsilon_{\text{eff}})\nabla E^2}{2} + \rho_{ext}\mathbf{E}$$

$$\tag{31}$$

where the external charge density is give by div $\mathbf{D} = \rho_{ext}$. This charge density is the overall external charge density in the dielectric, which may have been supplied by the battery, electrodes, and the surrounding air. In Eq.(31), the pressure P_o (ρ,T) is that which would be present in the absence of the electric field. In the case of a linear medium, the dielectric function ε is independent of field E, and $\varepsilon_{eff} = \varepsilon$, which reduces to the result derived by Landau and Lifshitz (see their Eq. (15.12) in Ref.[4]).

The total force on the fluid dielectric, $\mathbf{F}_{dielectric}$, is given by the volume integral of f over the volume of the dielectric Ω:

$$\mathbf{F}_{dielectric} = \int_{\Omega} f \, dV \qquad (32)$$

The volume Ω is the whole volume outside the metal electrodes of the capacitor. According to Equation (24), the net force on the capacitor $\mathbf{F}_{electrodes}$ is the negative of the total force on the dielectric:

$$\mathbf{F}_{electrodes} = \int_{\Omega} \left\{ \frac{E^2 \nabla \varepsilon}{2} + \frac{\nabla \left[(\varepsilon_{eff} - \varepsilon) E^2 \right]}{2} - \nabla \left[\frac{E^2 \rho}{2} \left(\frac{\partial \varepsilon_{eff}}{\partial \rho} \right)_{T,E} \right] - \rho_{ext} \mathbf{E} \right\} dV$$

$$(33)$$

where we have dropped the term containing the gradient in the pressure, assuming that it is negligible. Equation (33) gives the net force on capacitor plates for the case where the fluid dielectric is nonlinear, having the response given in Eq. (27). In Eq. (33), both dielectric constants are functions of the electric field. Note that the first three terms of the integrand depend on the square of the electric field, which is in agreement with the fact that the observed force direction is independent of the polarity of the applied bias.

There are four terms in the force. The first term is proportional the gradient of the dielectric constant, $\blacktriangledown \varepsilon$. We expect that the dielectric constant has a large variation in between regions of low and high electric field, such as near the smaller electrode. We expect that there is a strong nonlinear dielectric response due to ionization of the air. The resulting charges can move large distances, leading to a highly nonlinear response at high electric fields. Therefore, it is possible that this first term in the integrand in Eq. (33) has the dominant contribution. We expect this term to contribute to a force that point toward the smaller electrode (as observed experimentally), and we expect that this contribution is nearly independent of polarity of applied bias (except for asymmetric plasma effects under change of polarity).

The second term in the force Eq. (33) is proportional to the gradient of the product of the square of the electric field and the difference in dielectric constants. The difference in the dielectric constants, ε_{eff} - ε, can be expanded in a Taylor series in E

$$\varepsilon_{eff} - \varepsilon = -\frac{1}{3}\varepsilon'(0)E - \frac{1}{4}\varepsilon''(0)E^2 + ... \tag{34}$$

where

$$\varepsilon'(0) = \left(\frac{\partial \varepsilon}{\partial E}\right)_{T,\rho,E=0} \quad \text{and} \quad \varepsilon''(0) = \left(\frac{\partial^2 \varepsilon}{\partial E^2}\right)_{T,\rho,E=0} \tag{35}$$

The gradient of the square of the electric field always points toward the smaller electrode, independent of the polarity of bias applied to the capacitor. We do not know the sign of the dielectric constants $\varepsilon'(0)$ and $\varepsilon''(0)$ If the air has dielectric properties described by $\varepsilon'(0) < 0$ and $\varepsilon''(0) < 0$, then this term would contribute to a force toward the smaller electrode (which would be in agreement with experiment).

Alternatively, the term $\quad \frac{1}{2}\nabla\left[(\varepsilon_{eff} - \varepsilon)E^2\right]$

may have the wrong sign but may be small. This must be determined experimentally by studying the dielectric properties of air or other gas. The third term in the force Eq. (33) is difficult to evaluate. It may well be negligible, especially compared to the first term (assuming highly nonlinear dielectric response at high fields). Alternatively, if the air behaves as a nearly linear dielectric medium, then $\varepsilon_{eff} \approx \varepsilon$, and the dielectric constant of a gas is typically proportional to its density, $\varepsilon = \alpha \varepsilon_o \rho$, where ε_o is the permittivity of free space, and α is a constant. Using these expressions in Eq. (33) for ε yields the force on the capacitor electrodes for the case of a linear dielectric fluid:

$$\left(\mathbf{F}_{electrodes}\right)_{Linear\ Medium} = \int_{\Omega}\left\{-\frac{1}{2}\varepsilon\nabla E^2 - \rho_{ext}\mathbf{E}\right\}dV \tag{36}$$

50

For a linear medium, the first term in Eq. (35) contributes to a force pointing in a direction that is opposite to the gradient of the square of the electric field, i.e., it points toward the larger electrode (opposite to the experimentally observed force). In order to obtain a net force from Eq. (35) that is oriented toward the smaller electrode, the second term in Eq. (35) would have to dominate, i.e., the net force on the capacitor would be due to external charge effects. The magnitude of the external charges (from battery and surrounding air) on the dielectric fluid is unknown and must be determined experimentally. If the space between the capacitor plates is filled with a vacuum instead of dielectric, Eq. (33) reduces to a force given by

$$\left(\mathbf{F}_{electrodes}\right)_{Vacuum} = -\int_{\Omega} \rho_{ext} \mathbf{E} \quad dV \tag{37}$$

where the charge density ρ_{ext} is a complicated quantity, due to emission of electrode material and free charges, such as exists in a vacuum tube. The magnitude and sign of the force cannot be determined until a calculation is done of the charge density, ρ_{ext}, and electric field, $\mathbf{E}$.

The thermodynamic theory presented here provides a general expression in Eq. (33) for the net force on a capacitor in terms of the macroscopic electric field $\mathbf{E}$. This electric field in Eq. (33) must be determined by a microscopic calculation, taking into account the ionization of gas between capacitor plates, and details of charge transport.

In summary, at the present time, the relative magnitudes of the fours terms in the force expression given in Eq. (33) are unknown. The magnitudes of these terms must be determined by constructing a set of experiments designed to determine the field-dependent dielectric properties of the fluid (given by ε) surrounding the asymmetric capacitor electrodes. These experiments will permit us to verify if the thermodynamic theory presented here can explain the magnitude and sign of the observed force.

6. Summary and Suggested Future Work

We have presented a brief history of the Biefeld-Brown effect: a net force is observed on an asymmetric capacitor when a high voltage bias is applied. The physical mechanism responsible for this effect is unknown. In Section 4, we have presented estimates of the force on the capacitor due to the effect of an ionic wind and due to charge drift between capacitor electrodes. The force due to ionic wind is at least three orders of magnitude too small. The force due to the effect of charge drift is plausible, however, the estimates are only scaling estimates, not a microscopic model.

In Section 5, we have presented a detailed thermodynamic theory of the net force on a capacitor that is immersed in a nonlinear dielectric fluid, such as air in a high electric field. The main result for the net force on the capacitor is given in Eq. (33). The thermodynamic theory requires knowledge of the dielectric properties of the gas surrounding the capacitor plates.

It is not possible to estimate the various contributions to the force until we have detailed knowledge about the high-field dielectric properties of the fluid.

Much more experimental and theoretical work is needed to gain an understanding of the Biefeld-Brown effect. As discussed above, the most pressing question is whether the Biefeld-Brown effect occurs in vacuum. It seems that Brown may have tested the effect in vacuum, but not reported it, see Appendix B. More recently, there is some preliminary work that tested the effect in vacuum, and claimed that there is some small effect—smaller than the force observed in air, see the second report cited in Ref. [2]. Further work must be done to understand the effect in detail.

A set of experiments must be performed in vacuum, and at various gas pressures, to determine the actual facts about the effect. A careful study must be made of the force as a function of applied voltage, gas species, and gas pressure. In light of the thermodynamic theory presented here, the dielectric properties of the gas used in the experiments must be carefully measured. Obtaining such data will be a big step toward developing a theoretical explanation of the effect. On the theoretical side, a microscopic model of the capacitor (for a given geometry) must be constructed, taking into account the complex physics of ionization of air (or other gas) in the presence of high electric fields. Only by understanding the Biefeld-Brown effect in detail can its potential for applications be evaluated.

Acknowledgments

One of the authors (T.B.B.) thanks W. C. McCorkle, Director of Aviation and Missile Command, for the suggestion to look at the physics responsible for the net force on an asymmetric capacitor. The authors want to thank Jean-Louis Naudin (JLN Labs) for his permission to reproduce the letter of Thomas Townsend Brown in Appendix B. One of the authors (T.B.B.) is grateful for personal correspondence with Jean-Louis Naudin (JLN Labs).

Appendix A: Short Patent History Dealing with Asymmetric Capacitors[39]

T. Townsend Brown, "A method of and an apparatus or machine for producing force or motion", GB Patent N°300,311 issued on Nov 15, 1928[6]

Thomas Townsend Brown, "Electrokinetic Apparatus", US Patent N°2,949,550 issued on Aug 16, 1960

A.H. Bahnson Jr., "Electrical thrust producing device", US Patent N°2,958,790 issued on Nov 1, 1960

Thomas Townsend Brown, "Electrokinetic Transducer", US Patent N°3,018,394 issued on Jan 23, 1962

Thomas Townsend Brown, "Electrokinetic Generator", US Patent N°3,022,430 issued on Feb 20, 1962

Thomas Townsend Brown, "Electrokinetic Apparatus", US Patent N°3,187,206 issued on Jun 1, 1965

A.H. Bahnson Jr., "Electrical thrust producing device", US Patent N°3,223,038 issued on Dec 14, 1965

A.H. Bahnson Jr., "Electrical thrust producing device", US Patent N°3,227,901 issued on Jan 4, 1966

A.H. Bahnson Jr., "Electrical thrust producing device", US Patent N°3,263,102 issued on July 26, 1966

---

[39] This patent list was expanded to ensure completeness but those like Brown's 2,417,347 "vibration damper" are omitted. Several patent cover pages are displayed in the Patent Section on p.146 – Ed note

Thomas Townsend Brown, "Electrohydrodynamic Fluid Pump", US Patent N°3,267,860 issued on Aug 23, 1966

Hector Serrano (Gravitec), "Propulsion device and method employing electric fields for producing thrust" PCT Patent No. WO 00/58623, issued on October 5, 2000

Jonathan W. Campbell, (NASA), "Apparatus for generating thrust using a two dimensional, asymmetrical capacitor module" US Patent No. 6,317,310 issued on Nov. 13, 2001

Jonathan W. Campbell, (NASA), "Apparatus for generating thrust using a two dimensional, asymmetrical capacitor module" US Patent No. 6,411,493 issued on January 31, 2002

Jonathan W. Campbell, (NASA), "Apparatus for Generating Thrust using a two dimensional asymmetric capacitor module" Patent No. 6,411,493 issued on June 25, 2002

Jonathan W. Campbell, (NASA), "Cylindrical asymmetrical capacitor devices for space applications" US Patent No. 6,775,123 issued on Aug. 10, 2004

Appendix B: Force on Asymmetric Capacitor in Vacuum
 Enclosed below is a copy of my email correspondence with Jean-Louis Naudin, an expert in Lifters, which hosts a web site on the subject and is the moderator of a yahoo.com newsgroup named "Lifters". In this correspondence, Naudin quotes a letter, purportedly signed by T. Townsend Brown, in which Brown discusses the question of whether an asymmetric capacitor has a net force on it in vacuum under high voltage.

Subject : Re Do Lifters Work in Vacuum?
From: tbahder@att.net
Date: Sun Sep 15, 2002 5:50am
Subject: Do Lifters work in vacuum?

I am new to this effect. However, I have constructed a standard Lifter, and have confirmed that it produces the same apparent lift under D.C. under both polarities. That is a really weird result because everything that I can think of in classical electromagnetic theory has a polarity dependence (for D.C. effects). I am trying to understand the physics of the lifter. The next question is whether a lifter will work in vacuum. I thought that several days ago I came across a web site that showed an experiment that demonstrated the lifter operating in vacuum. However, that was a few days ago, and I cannot locate the web site now. Was it my imagination, or has operation in vacuum been demonstrated?

Thanks,
Tom Bahder, Clarksville, MD U.S.A.

Those who have been experimenting with electrokinetic thrusters such as the Lifter admit that part of the effect is due to ionic wind, but that there is another effect that would still operate in a vacuum.

Now Gravitec has a report on their website from Purdue University, Energy Conversion Lab. See:
http://foldedspace.com/EKP%20Ionic%20Wind%20Study%20-%20Purdue.doc

The purpose of the report is to explore the possibility that Electrokinetic Propulsion is just another manifestation of the Ionic Wind Effect. Three different cases were explored; the first being normal atmospheric operation, in which the surrounding atmosphere was ionized. The second case used atmospheric ions present within a vacuum. The last case used the actual dielectric media as the ions. Also the expected theoretical result in vacuum is off by a factor of more than a thousand (being a thrust of 3.65 e-4 mN expected, whereas a force of at least .31mN was observed at a lower voltage of 17kV). These were the only observations recorded, since it was deemed unnecessary to try to take readings within a vacuum since the observed and experimental currents are off by orders of magnitude and not enough to produce any meaningful effect during Electrokinetic Propulsion experiments.

Gravitec admits the following: The initial vacuum test showed as suspected that field propulsion did not require any exhaust gasses to operate. These tests, while good, are not enough to bring to the scientific community, because something this extraordinary in nature needs extraordinary proof. We currently need to perform a more controlled and metered vacuum experiment to eliminate all doubts that have surrounded the phenomenon in the past. It is our desire that the new vacuum test be done at one of the vacuum facilities at the Naval Research Laboratories (NRL) in Washington DC, a NASA testing facility or in some other well equipped French labs sometime in the near future. The test and the set ups will be created by Gravitec, in co-operation with the testing facility. There will also be other participants including Dr. John Rusek and members of various interested government agencies.

You will find below a copy of a letter from The Townsend Brown Foundation, Ltd. Nassau, Bahamas and dated February 14, 1973 arrived, carrying the following information, personally signed by T.Townsend Brown.

Dear ....,

You have asked several question which I shall try to answer. The experiments in vacuum were conducted at "Societe Nationale de Construction Aeronautique" in Paris in 1955-56, in the Bahnson Laboratories, Winston- Salem, North Carolina in 1957-58 and at the "General Electric Space Center" at King of Prussia, Penna, in 1959. Laboratory notes were made, but these notes were never published and are not available to me now. The results were varied, depending upon the purpose of the experiment. We were aware that the thrust on the electrode structures were caused largely by ambient ion momentum transfer when the experiments were conducted in air. Many of the tests, therefore, were directed to the exploration of this component of the total thrust. In the case of the G.E. test, cesium ions were seeded into the environment and the additional thrust due to seeding was observed. In the Paris test miniature saucer type airfoils were operated in a vacuum exceeding 10^{-6}mm Hg. Bursts of thrust (towards the positive) were observed every time there was a vaccum spark within the large bell jar. These vacuum sparks represented momentary ionization, principally of the metal ions in the electrode material. The DC potential used ranged from 70kV to 220kV. Condensers of various types, air dielectric and barium titanate were assembled on a rotary support to eliminate the electrostatic effect of chamber walls and observations were made of the rate of rotation. Intense acceleration was always observed during the vacuum spark (which, incidentally, illuminated the entire interior of the vacuum chamber). Barium Titanate dielectrique always exceeded air dielectric in total thrust. The results which were most significant from the standpoint of the Biefeld-Brown effect was that thrust continued, even when there was no vacuum spark, causing the rotor to accelerate in the negative to positive direction to the point where voltage had to be reduced or the experiment discontinued because of the danger that the *rotor would fly apart. In short, it appears there is strong evidence that Biefeld-Brown effect does exist in the negative to positive direction in a vacuum of at least 10^{-6} Torr. The residual thrust is several orders of magnitude larger than the remaining ambient ionization can account for. Going further in your letter of January 28th, the condenser "Gravitor" as described in my British patent, only showed a loss of weight when vertically oriented so that the negative-to-positive thrust was upward. In other words, the thrust tended to "lift" the gravitor. Maximum thrust observed in 1928 for one gravitor weighing approximately 10 kilograms was 100 kilodynes at 150kV DC. These gravitors were very heavy, many of them made with a molded dielectric of lead monoxide and beeswax and encased in bakelite. None of these units ever "floated" in the air. There were two methods of testing, either as a pendulum, in which the angle of rise against gravity was measured and charted against the applied voltage, or, as a rotor 4ft. in diameter, on which four "gravitors" were mounted on the periphery. This 4 ft. wheel was tested in air and also under transformer oil. The total thrust or torque remained virtually the same in both instances, seeming to prove that aero-ionization was not wholly responsible for the thrust observed. Voltage used on the experiments under oil could be increased to about 300kV DC and the thrust appeared to be linear with voltage.*

In subsequent years, from 1930 to 1955, critical experiments were performed at the Naval Research Laboratory, Washington, DC.; the Randall-Morgan Laboratory of Physics, University of Penna., Philadelphia; at a field station in Zanesville, Ohio, and two field stations in Southern California, of the torque was measured continuously day and night for many years. Large magnitude variations were consistently observed under carefully controlled conditions of constant voltage, temperature, under oil, in magnetic and electrostatic shields, not only underground but at various elevations. These variations, recorded automatically on tape, were statistically processed and several significant facts were revealed. There were pronounced correlations with mean solar time, sidereal time and lunar hour angle. This seemed to prove beyond a doubt that the thrust of "gravitors" varied with time in a way that related to solar and lunar tides and sidereal correlation of unknown origin. These automatic records, acquired in so many different locations over such a long period of time, appear to indicate that the electrogravitic coupling is subject to an extraterrestrial factor, possibly related to the universal gravitational potential or some other (as yet) unidentified cosmic variable. In response to additional questions, a reply of T.T.

Brown, dated April, 1973, stated "The apparatus which lifted itself and floated in the air, which was described by Mr Kitselman, was not a massive dielectric as described in the English patent. Mr Kitselman witnessed an experiment utilizing a 15" circular, dome-shaped aluminum electrode, wired and energized as in the attached sketch. When the high voltage was applied, this device, although tethered by wires from the high voltage equipment, did rise in the air, lifting not only its own weight but also a small balance weight which was attached to it on the underside. It is true that this apparatus would exert a force upward of 110% of its weight. The above experiment was an improvement on the experiment performed in Paris in 1955 and 1956 on disc air foils. The Paris experiments were the same as those shown to Admiral Radford in Pearl Harbor in 1950. These experiments were explained by scientific community as due entirely to "ion-momentum transfer", or "electric wind". It was predicted categorically by many "would-be" authorities that such an apparatus would not operate in vacuum. The Navy rejected the research proposal (for further research) for this reason. The experiments performed in Paris several years later, proved that ion wind was not entirely responsible for the observed motion and proved quite conclusively that the apparatus would indeed operate in high vacuum. Later these effects were confirmed in a laboratory at Winston-Salem, N.C., especially constructed for this purpose. Again continuous force was observed when the ionization in the medium surrounding the apparatus was virtually nil. In reviewing my letter of April 5th, I notice, in the drawing which I attached, that I specified the power supply to be 50kV. Actually, I should have indicated that it was 50 to 250kV DC for the reason that the experiments were conducted throughout that entire range. The higher the voltage, the greater was the force observed. It appeared that, in these rough tests, that the increase in force was approximately linear with voltage. In vacuum the same test was carried on with a canopy electrode approximately 6" in diameter, with substantial force being displayed at 150 kV DC. I have a short trip of movie film showing this motion within the vacuum chamber as the potential is applied."
Kindest personal regards,

Sincerely,
T..Townsend Brown

References

- There are numerous references to asymmetric capacitors, called "lifters" on the internet, see web sites: http://jnaudin.free.fr/ (JNaudin)
- http://www.soteria.com/brown/ (Web site summarizing information about Thomas Townsend Brown,
- http://www.tdimension.com/(TransdimensionalTechnologies)
- http://www.jlnlabs.org (J. Naudin)
- http://tventura.hypermart.net/index.html (American Antigravity).
- William B. Stein, "Electrokinetic Propulsion: The Ionic Wind Argument", Purdue University Energy Conversion Lab, Hangar #3, Purdue Airport West Lafayette, IN 47906, Sept. 5, 2000, on web at http://foldedspace.com/EKP%20Ionic%20Wind%20Study%20-%20Purdue.doc.
- R. L. Talley, "Twenty First Century Propulsion Concept", Veritay Technology, Inc. 4845 Millersport Highway, East Amherst, N.Y. 14051, Report prepared for the Phillips Laboratory,Air Force Systems Command, Propulsion Directorate, Edwards AFB CA 93523- 5000.
- See the web site EvgenijBarsoukov,http://sudy_zhenja.tripod.com/lifter_theory /. See sections 2, 5 and 15 of L. D. Landau and E. M. Lifshitz, "Electrodynamics of Continuous Media", 2nd Edition, Pergamon Press, N.Y. 1984.
- See p. 191 in L. B. Loeb, "Fundamental Processes of Electrical Discharges in Gases", John Wiley and Sons, N.Y., 1939.
- See p. 62 in S. C. Brown, "Basic Data of Plasma Physics", John Wiley and Sons, N.Y. 1959.
- See p. 139, J. A. Stratton, "Electromagnetic Theory", McGraw Hill Book Company, N. Y.,1941.
- See p. 95 in M. Abraham and R. Becker, "The Classical Theory of Electricity and Magnetism Hafner Publishing Co. Inc., N. Y., Second Edition, 1950.
- See p. 104, J. A. Stratton, "Electromagnetic Theory", McGraw Hill Book Company, N. Y.,1941.

Possibility of a Strong Coupling Between Electricity and Gravitation

Takaaki Musha

3-11-7-601 Namiki, Kanazawa-ku, Yokohama 236-0005 Japan

E-mail: musha@jg.ejnet.ne.jp

Abstract

The finding of Prof. Biefeld and T.T. Brown, which is called the Biefeld-Brown effect, suggests strong coupling between electricity and gravitation. However this phenomenon can not be predicted within the framework of conventional physics. The author attempts to explain this phenomenon by introducing an asymmetrical gravitational field generated inside the atom by a high potential electric field; he also verifies the theoretical value compared with the experimental result.

Introduction

Prof. Biefeld and T.T. Brown discovered that a high potential charged capacitor with dielectrics exhibited unidirectional thrust toward the positive plate when the atoms of a material are placed within the electric field of a capacitor. This phenomenon is called the Biefeld-Brown effect (B-B effect) and it suggests a connection between electricity and gravitation.

Characteristics of the B-B effect can be summarized shown as follows:[1]

- The separation of the plates of the condenser—closer plates, greater effect.
- The higher the specific inductive capacity of the dielectrics between the plates, the greater the effect.
- The greater the area of the condenser plates, the greater the effect.
- The greater the voltage difference between the plates, the greater the effect.
- The greater the mass of the material between the plates, the greater the effect.

However, the coupling between electrostatic and gravitational fields can be predicted neither by General Relativity, nor conventional field theory. The author attempts to explain this phenomenon by

60

introducing a new gravitational field generated inside the atom by a high potential electric field.

Theoretical Consideration on the B-B Effect

Weak field approximation of Einstein's General Relativity leads to the generalized formula of Lorenz force given by[2]

$$F = q(E + v \times B) + m(E_g + v \times B_g),$$ (1)

where q is the charge of the particle, m is the mass of the particle, E is the electric field, B is the magnetic field, v is the velocity of the particle, E_g is the electrogravitic field, and B_g is the magnetogravitic field.

From which, gravitoelectric-electric coupling inside the static atom under high electric potential field becomes

$$qE + mE_g = 0,$$ (2)

by assuming that the internal volume of an elementary particle is a region of force-free field like a superconductor.[3] Then the gravitational field generated at the center of the charged particle by an external electric field becomes

$$E_g = -(q/m)E.$$ (3)

Comparing q/m values of an electron and a pion, E_g can be generated by an electron rather than a pion, hence we can let $q \approx e$ and $m \approx m_e$, where e is a charge of an electron and m_e is its mass. For the estimation of gravitational effect, we introduce the following approximation of the electrogravitic potential given by

$$\Phi_g = -(e/m_e)E\{\delta^2 x/(\delta^2 + x^2)\},$$ (4)

which satisfies following conditions:

$$\partial\Phi_g(0)/\partial x = -(e/m_e)E,$$ (5.1)

$$\partial\Phi_g(\pm\infty)/\partial x = 0,$$ (5.2)

where d is a length of the domain at which the new gravitational field is generated.

Figure 1 shows the deformation of the space generated at the center of the elementary particle and the shape of the electrogravitic potential. By the asymmetric electron orbit shown in Figure 2 generated by the external electric field, the electrogravitic potential at the center of the atomic nucleus becomes

61

$$\Phi_g = \Phi_g(r + \lambda) + \Phi_g(r - \lambda), \qquad (6)$$

where λ is a displacement of charge by an applied electric field and r is an orbital radius of the electron around the nucleus.

From which, the electrogravitic force generated by electrons circulating around the nucleus, the number of them equals the atomic number Z, is given by [4,5]

$$E_g = \partial\Phi_g / \partial x = \delta^2 eE / m_e \sum_{i=1}^{Z}[(r_i + \lambda)^{-2} + (r_i - \lambda)^{-2}], \qquad (7)$$

when satisfying $|r_i \pm \lambda| \gg \delta$, where r_i is the orbital radius of each electron around the atomic nucleus. For simplifying the problem, we set $r_i \approx r_0$, then Equation 7 becomes

$$E_g \approx Z\delta^2 eE / m_e[(r_0 + \lambda)^{-2} + (r_0 - \lambda)^{-2}]. \qquad (8)$$

For relative lower voltage, we can suppose that $r_0 \gg \lambda$. Then Equation 8 can be approximated as

$$E_g \approx -2\delta^2 Ze /(m_e r_0^2)\{1 + 3\lambda^2 / r_0^2 + \cdots\}E. \qquad (9)$$

If we suppose that additional equivalent mass in a space due to the electric field[6] is canceled by negative mass created by the new gravitational field, we have the following formula given by

$$\int_V 2\pi G\varepsilon E^2 / c^2 dv - \int_V E_g^2 /(2c^2)dv = 0, \qquad (10)$$

where ε is permittivity, G is a gravitational constant, and c is a light speed. From which, we have

$$\delta \approx \sqrt[4]{\pi\varepsilon G m_e^2 r_0^4 / e^2}. \qquad (11)$$

If r_0 is replaced by Bohr's radius and ε equals ε_0, which is the permittivity of free space, we obtain the length of a domain where the new gravitational field is generated to be $\delta \approx 8.3 \times 10^{-22}$ (m), that is much smaller than the radius of electron, which is about 2.8×10^{-15} (m). Then acceleration of the dielectric material induced by high potential electric field can be approximated from Equations 9 and 11 as

$$E_g \approx -Z\sqrt{4\pi\varepsilon_r\varepsilon_0 G} \cdot E = -8.62 \times 10^{-11} Z\sqrt{\varepsilon_r} \cdot E, \qquad (12)$$

where ε_r is the specific inductive capacity of the dielectric material determined as $\varepsilon = \varepsilon_r \varepsilon_0$. From which, the force generated by high potential electric field becomes

$$F \approx 8.62 \times 10^{-11} Z\mu S \sqrt{\varepsilon_r} \cdot V / t, \qquad (13)$$

where μ is a total mass of the dielectric material per unit area, S is an area of the capacitor, t is a thickness of the capacitor, and V is an impressed voltage.

This equation presented here satisfies characteristics 1 through 5 mentioned in the previous section and it also predicts another characteristic for the B-B effect shown as follows:

The greater the atomic number of the material between the plates, the greater the effect

Experimental Results

From February 1 until March 1, 1996, a research group at the HONDA R&D Institute in Japan conducted experiments to verify the B-B effect with an improved experimental device[7,8] to reject the influence of corona discharges and electric wind around the capacitor. For the rejection of these effects, the capacitor was set in the insulator oil contained within a metallic vessel as shown in Figure 3. The capacitor used at the experiment (Figure 4) was a circular plate made of glass with thickness=1mm, diameter=170mm and weight=62g. The specific inductive capacity of the capacitor was about 10. They conducted experiments for both cases, +DC 18kv and -DC 18kv applied to the capacitor. The experimental results obtained by the HONDA research group are shown in Figure 5, where the horizontal line is for the number of the experiment which they conducted and the vertical line is for the weight loss of the capacitor.

Analysis of the Experimental Results

By the experiment results, the coefficient κ as follows can be determined by

$$\kappa = E_g / E = (\Delta W / W)(t / V)g_0, \qquad (14)$$

where ΔW is a weight loss of the capacitor under high potential electric field, W is a weight of the capacitor, and g_0 is the gravitational acceleration at the Earth's surface.

From which, histograms of the experiment were obtained, as shown in Figure 7, where the horizontal line shows the value K, which equals

63

κ times 10^9, and the vertical line shows the number of observed results. Statistical analysis shows that the mean value of K obtained is 2.59 and its standard deviation is 1.46, then we have $\kappa \approx (2.59 \pm 1.46) \times 10^{-9}$.

From Equation 12, the absolute value of κ can be given by

$$|\kappa| = 8.62 \times 10^{-11} Z \sqrt{\varepsilon_r} . \qquad (15)$$

Most of ingredient of glass is S_iO_2, the mean value of Z per unit atom can be roughly estimated as $Z \approx (14 + 8 \times 2)/3 = 10$, so we have $|\kappa| \approx 2.73 \times 10^{-9}$.

From which, we have

$$(E_g)_{measured} / (E_g)_{theoretical} = 0.95 \pm 0.53 . \qquad (16)$$

Thus it is considered that the force generated due to the external electric field can be a new gravitational field generated in the micro level by strong coupling between electromagnetism and gravitation. By using the equation which the author derived, Iwanaga analyzed the effectiveness of the B-B effect as a propulsion system in his paper[9] and he estimated the thrust obtained by the B-B effect is much better than that of an arc-jet propulsion system, as shown in Table 1. He considered it would be profitable to apply this propulsion method to small space vehicles.

Table 1 Thrust of the propulsion systems and B-B effect.

Propulsion System	Thrust (N)
Jet engine	2×10^5
Chemical fuel rocket	2.45×10^5
Arc-jet	1.50×10^{-1}
Nuclear power	8.82×10^5
Photon rocket	3.3
B-B effect[*]	100

---

[*] Calculated for mass=100Kg, diameter/height ratio=100, $\varepsilon_r = 5$ and $E = 7 \times 10^8$ V/m [Reference 9].

Conclusion

The theoretical equation for the B-B effect, which satisfies the characteristics clarified by T. T. Brown, is derived. Because the experimental results agree well with the theoretical calculation, it is considered that the force generated due to the external electric field is due to a new gravitational field generated at the micro level induced by strong coupling between electromagnetism and gravitation.

References

1. Sigma,Rho.1986. Ether-Technology: A Rational Approach to Gravity Control, Cadake Industries, Clayton. GA

2. Harris,E.G.1991. "Analogy Between General Relativity and Electromagnetism for Slowly Moving Particles in Weak Gravitational Fields." American Journal of Physics,59, 5,421-425.

3. Torr,D.G. and Li,N.1993. "Gravitoelectric-electric Coupling via Superconductivity," Foundations of Physics Letters, 6, 4,371-383.

4. Musha,T. and Sawatari,K.1992. "Possibility of Space Drive Propulsion by High Potential Field," Proceedings of the 36th Space Science and Technology Conference, JSASS,95-96(in Japanese).

5. Musha,T. and Abe,I.1993. "Biefeld-Brown Effect and Electro-gravitic Propulsion by High Potential Electric Field," Proceedings of the 24th JSASS Annual Meeting, JSASS, 189-192 (in Japanese).

6. Brillouin,L.1970. Relativity Reexamined, Academic Press.Inc., New York.

7. Musha,T.1996. "Brown's Electro-Gravitic Propulsion System",Sec.3.6.3,Report of Advanced Space Propulsion Investigation Committee, JSASS,104-116 (in Japanese).

8. Musha,T.2000. "Theoretical Explanation of the Biefeld-Brown Effect," Electric Spacecraft Journal, 31, 29.

9. Iwanaga,N.1999. "Review of Some Field Propulsion Methods from the General Relativistic Standpoint," Space Technology and Applications International Forum," American Institute of Physics, 1051-1059.

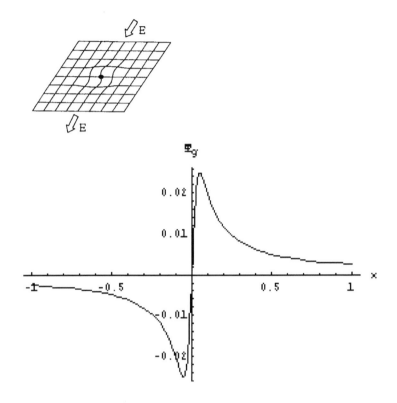

Figure 1. New gravitational field generated by an external electric field.

Figure 2. Deformation of the atom in an external electric field.

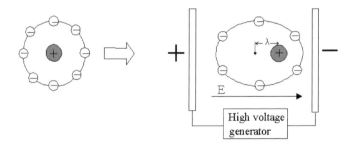

High voltage generator

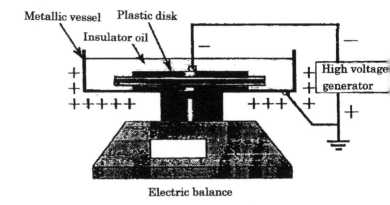

Figure 3. Experimental set-up.

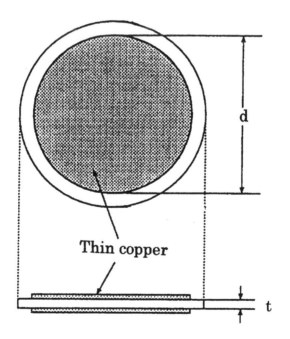

Figure 4. Capacitor used for the experiment.

67

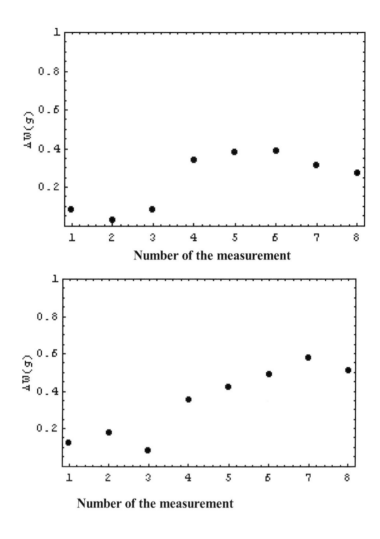

Figure 5. Weight reduction observed at the improved experiment
(upper figure: +18kV, lower figure: -18kV).

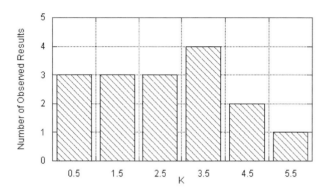

Figure 6. Histogram of the experimental results.

HISTORICAL SECTION

Bahnson holding "experimental ballistic electrode"

How I Control Gravitation
T. Townsend Brown, 1929

Published in *Science and Invention*, August 1929, and
Psychic Observer, Vol. XXXVII, No. 1, 1929

There is a decided tendency in the physical sciences to unify the great basic laws and to relate, by a single structure or mechanism, such individual phenomena as gravitation, electrodynamics and even matter itself.

It is found that matter and electricity are very closely related I structure. In the final analysis, matter loses its traditional individuality and becomes merely an "electrical condition." In fact, it might be said that the concrete body of the universe is nothing more than an assemblage of energy, which in itself, is quite intangible.

Of course, it is self evident that matter is connected with gravitation and it follows logically that electricity is likewise connected. These relations exist in the realm of pure energy and consequently are very basic in nature. In all reality, *they constitute the true backbone of the universe.*

It is needless to say that the relations are not simple, and full understanding of their concepts is complicated by the outstanding lack of information and research on the real nature of gravitation. The theory of relativity introduce a new and revolutionary light to the subject by injecting a new conception of space and time. Gravitation thus becomes the natural outcome of so called "distorted space." It loses its Newtonian interpretation as a tangible mechanical force and gains the rank of an "apparent" force, due merely to the condition of space itself.

Fields in space are produced by the presence of material bodies or electric charges. They are gravitational fields or electric fields according to their causes. Apparently, they have no connection, one with the other. This fact is substantiated by observations to the effect that electric fields can be shielded and annulled, while gravitational fields are nearly perfectly penetrating. This dissimilarity has been the chief hardship to those who would compose a theory of combination.

It required Dr. Einstein's own close study for a period of several years to achieve the results others have sought in vain and to announce with certainty the unitary field laws.

Einstein's field theory is purely mathematical. It is not based on the results of any laboratory test and does not, so far as known, predict any method by which an actual demonstration or proof may be made. The new theory accomplishes its purpose by "rounding out" the accepted principles of relativity so as to embrace electrical phenomena.

The theory of relativity, thus supplemented, represents the last word in mathematical physics. It is most certainly a theoretical structure of overpowering magnitude and importance. The thought involved is so far-reaching that it may be many years before the work is fully appreciated and understood.

Early Investigations

The writer and his colleagues anticipated the present situation even as early as 1923, and began at that time to construct the necessary theoretical bridge between the two then separate phenomena, electricity and gravitation.

The first actual demonstration of the relation was made in 1924. Observations were made of the individual and combined motions of two heavy lead balls which were suspended by wires 45 cm apart. The balls were given opposite electrical charges and the charges were maintained.

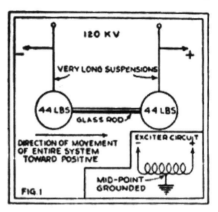

A SIMPLE TYPE OF GRAVITATOR IS SHOWN IN THE ABOVE ILLUSTRATION.

Sensitive optical methods were employed in measuring the movements and, as near as could be observed, the balls appeared to behave according to the following law: "Any system of two bodies possesses a mutual and unidirectional force (typically in the line of the bodies) which is directly proportional to the product of the masses, directional proportional to the potential difference and inversely proportional to the square of the distance between them."

$$F \approx \frac{Vm_1 m_2}{r^2}$$

The peculiar result is that the gravitational field of the earth had no apparent connection with the experiment. The gravitational factors entered through the consideration of the mass of the electrified bodies.

72

The newly discovered force was quite obviously the resultant physical effect of an electrogravitational interaction. It represented the first

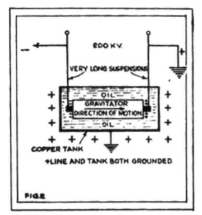

actual evidence of the very basic relationship. The force was named "gravitator action," for want of a better term, and the apparatus or system of masses employed was called a "gravitator" [see Figure 1].

Since the time of the first test, the apparatus and the methods used have been greatly improved and simplified. Cellular "gravitators" have taken the place of the large balls of lead. Rotating frames supporting two and four gravitators have made possible acceleration measurements. Molecular gravitators made of solid blocks of massive dielectric have given still greater efficiency. Rotors and pendulums operating under oil have eliminated atmospheric considerations as to pressure, temperature and humidity. The disturbing effects of ionization, electron emission and pure electrostatics have likewise been carefully analyzed and eliminated.

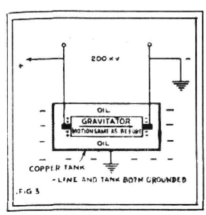

Finally after many years of tedious work and with refinement of methods we succeeded in observing the gravitational variations produced by the moon and sun and much smaller variations produced by the different planets. It is a curious fact that the effects are most pronounced when the affecting body is in the alignment of the differently charged elements and least pronounced when it is at right angles.

73

Much of the credit for this research is due to Dr. Paul Biefield, Director of Swazey Observatory. The writer is deeply indebted to him for his assistance and for his many valuable, timely suggestions.

Gravitator Action an Impulse

Let us take, for example, the case of a gravitator totally immersed in oil but suspended so as to act as a pendulum and swing along the line of its elements [see Figure 2]. When the direct current with high voltage (75-300 KV) is applied the gravitator swings up the arc until its propulsive force balances the force of the earth's gravity resolved to that point, then it stops, but it does not remain there. The pendulum then gradually returns to the vertical or starting position even while the potential is maintained. The pendulum swings only to one side of the vertical. Less than five seconds is required for the test pendulum to reach the maximum amplitude of the swing but from thirty to eighty seconds are required for it to return to zero [see Figure 3].

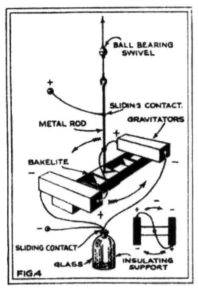

A GRAVITATOR ROTOR IS SIMPLY AN ASSEMBLY OF UNITS SO MADE THAT ROTATION RESULTS UNTIL THE IMPULSE IS EXHAUSTED.

The total time or duration of the impulse varies with such cosmic conditions as the relative position and distance of the moon, sun and so forth. It is in no way affected by fluctuations in the supplied voltage and averages the same for every mass or material under test. The duration of the impulse is governed solely by the condition of the gravitational field. It is a value which is unaffected by changes in the experimental set-up, voltage applied or type of gravitator employed. Any number of different kinds of gravitators operating simultaneously on widely different voltages would reveal exactly the same impulse duration at any instant. Over an extended period of time all gravitators would show equal variations in the duration of the impulse.

After the gravitator is once fully discharged, its impulse exhausted, the electrical potential must be removed for at least five minutes in

74

order that it may recharge itself and regain its normal gravitic condition. The effect is much like that of discharging and charging a storage battery, except that electricity is handled in a reverse manner. When the duration of the impulse is great the time required for complete recharge is likewise great. The times of discharge and recharge are always proportional.

Technically speaking, the *exogravitic rate* and the *endo-gravitic rate* are proportional to the *gravitic capacity*.

Summary of Characteristics

Summing up the observations of the electro-gravitic pendulum the following characteristics are noted [see Figures 4, 5, 6]:

- APPLIED VOLTAGE determines only the *amplitude* of the swing.

- APPLIED AMPERAGE is only sufficient to overcome leakage and maintain the required voltage through the losses of the dielectric. Thus the total load approximates on 37 ten-millionths of an ampere. It apparently has no other relation to the movement at least from the present state of physics.

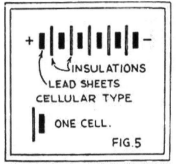

INSULATIONS
LEAD SHEETS
CELLULAR TYPE

ONE CELL.

FIG.5

THE CELLULAR GRAVITATOR IS BUILT IN THE FORM OF A HIGH VOLTAGE SERIES CONDENSOR.

- MASS OF THE DIELECTRIC is a factor in determining the total energy involved in the impulse. For a given amplitude an increase in mass is productive of an increase in the energy exhibited by the system ($E = mg$).

- DURATION OF THE IMPULSE with electrical conditions maintained is independent of all of the foregoing factors. It is governed solely by external gravitational conditions, positions of the moon, sun, etc., and represents the total energy or summation of energy values which are effective at that instant.

- GRAVITATIONAL ENERGY LEVELS are observable as the pendulum returns from the maximum deflection to the zero point or vertical position. The pendulum hesitates in its return

75

movement on definite levels or steps. The relative position and influence of these steps vary continuously every minute of the day. One step or energy value corresponds in effect to each cosmic body that is influencing the electrified mass or gravitator. By merely tracing a succession of values over a period of time a fairly intelligible record of the paths and the relative gravitational effects of the moon, sun, etc., may be obtained.

In general then, every material body possesses inherently within its substance separate and distinct energy levels corresponding to the gravitational influences of every other body. These levels are readily revealed as the electrogravitic impulse dies and as the total gravitic content of the body is slowly released.

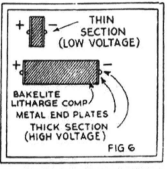

THE MOLECULAR TYPE GRAVITATOR IS MADE WITH A DIELECTRIC BLOCK AND METAL END PLATES OR ELECTRODES.

The gravitator, in all reality, is a very efficient electric motor. Unlike other forms of motors it does not in any way involve the principles of electromagnetism, but instead it utilizes the newer principles of electro-gravitation. A simple gravitator has no moving parts but is apparently capable of moving itself from within itself. it is highly efficient for the reason that it uses no gears, shafts, propellers or wheels in creating its motive power. It has no internal resistance and no observable rise in temperature. Contrary to the common belief that gravitational motors must necessarily be vertical-acting, the gravitator, it is found, acts equally well in every conceivable direction.

While the gravitator is at present primarily a scientific instrument, perhaps even an astronomical instrument, it also is rapidly advancing to a position of commercial value. Multi-impulse gravitators weighing hundreds of tons may propel the ocean liners of the future. Smaller and more concentrated units may propel automobiles and even airplanes. Perhaps even the fantastic "space cars" and the promised visit to Mars may be the final outcome. Who can tell?

Towards Flight – Without Stress or Strain or Weight

By Intel, Washington, D.C.

Interavia, March 23, 1956

The following article is by an American journalist who has long taken a keen interest in questions of theoretical physics and has been recommended to the Editors as having close connections with scientific circles in the United States. The subject is one of immediate interest and Interavia would welcome further comment from initiated sources. –Editors.

Electro-gravitics research, seeking the source of gravity and its control, has reached a stage where profound implications for the entire human race begin to emerge. Perhaps the most startling and immediate implications of all involve aircraft, guided missiles, atmospheric and free space flight of all kinds.

If only one of several lines of research achieve their goal and it now seems certain that this must occur, gravitational acceleration as a structural, aerodynamic and medical problem will simply cease to exist. So will the task of providing combustible fuels in massive volume in order to escape the earth's gravitic pull, now probably the biggest headache facing today's would-be "space men."

And towards the long-term progress of mankind and man's civilization, a whole new concept of electrophysics is being levered out into the light of human knowledge.

There are gravity research projects in every major country of the world. A few are over 30 years old.[40] Most are much newer. Some are purely theoretical and seek the answer in Quantum, Relativity and Unified Field Theory mathematics Institute for Advanced Study at Princeton, New Jersey; University of Indiana's School of Advanced Mathematical Studies; Purdue University Research Foundation; Goettingen and Hamburg Universities in France, Italy, Japan and elsewhere. The list, in fact, runs into the hundreds.

Some projects are mostly empirical, studying gravitic isotopes, electrical phenomena and the statistics of mass. Others combine both approaches in the study of matter in its super-cooled, super-conductive state, of jet electron streams, peculiar magnetic effects or the electrical

---

[40] Ultimately they go back to Einstein's general theory of relativity (1916), in which the law of gravitation was first mathematically formulated as a field theory, in contrast to Newton's "action-at-a-distance" concept. - Ed (*Interavia*)

mechanics of the atom's shell. Some of the companies involved in this phase include Lear Inc., Gluhareff Helicopter and Airplane Corp., The Glenn L. Martin Co., Sperry-Rand Corp., Bell Aircraft, Clarke Electronics Laboratories, the U.S. General Electric Company.

The concept of weightlessness in conventional materials which are normally heavy, like steel, aluminum, barium, etc., is difficult enough, but some theories, so far borne out empirically in the laboratory, postulate that not only can they be made weightless, but they can in fact be given a negative weight. That is: the force of gravity will be repulsive to them and they will, new sciences breed new words and meanings for old ones, loft away contra-gravitationally. In this particular line of research, the weights of some materials have already been cut as much as 30% by "energizing" them. Security prevents disclosure of what precisely is meant by "energizing" or in which country this work is under way. The American scientist T. Townsend

The American scientist Townsend T. Brown has been working on the problems of electrogravitics for more than thirty years. He is seen here demonstrating one of his laboratory instruments, a disc-shaped variant of the two-plate condenser.

Brown has been working on the problems of electrogravitics for more than thirty years. He is seen here demonstrating one of his laboratory instruments, a disc-shaped variant of the two-plate condenser. A

78

localized gravitic field used as a ponderomotive force has been created in the laboratory. Disc airfoils two feet in diameter and incorporating a variation of the simple two-plate electrical condenser charged with fifty kilovolts and a total continuous energy input of fifty watts have achieved a speed of seventeen feet per second in a circular air course twenty feet in diameter. More lately these discs have been increased in diameter to three feet and run in a fifty-foot diameter air course under a charge of a hundred and fifty kilovolts with results so impressive as to be highly classified. Variations of this work done under a vacuum have produced much greater efficiencies that can only described as startling. Work is now under way developing a flame-jet generator to supply power up to fifteen million volts.

Such a force rose exponentially to levels capable of pushing man-carrying vehicles through the air, or outer space, at ultrahigh speeds is now the object of concerted effort in several countries. Once achieved it will eliminate most of the structural difficulties now encountered in

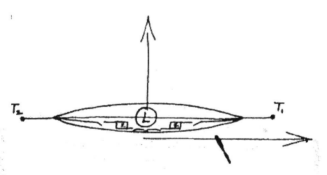

Author's sketch of a supersonic space ship roughly 50 ft. in diameter, whose lift and propulsion are produced by electrogravitic forces. The vehicle is supported by a " lofting cake " L consisting of " gravitic isotopes " of negative weight, and is moved in the horizontal plane by propulsion elements T, and T₁.

the construction of high-speed aircraft. Importantly the gravitic field that provides the basic propulsive force simultaneously reacts on all matter within that field's influence. The force is not a physical one acting initially at a specific point in the vehicle that needs then to be translated to all the other parts. It is an electrogravitic field acting on all parts simultaneously. Subject only to the so-far immutable laws of momentum, the vehicle would be able to change direction, accelerate to thousands of miles per hour, or stop. Changes in direction and

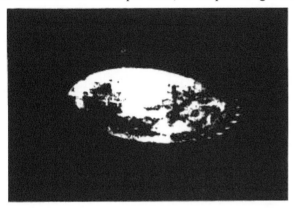

Townsend Brown's free-flying condenser. If the two arc-shaped electrodes (on the left and right rims) are placed under electrostatic charge, the disc will move, under the influence of interaction between electrical and gravitational fields, in the direction of the positive electrode. The higher the charge, the more marked will be the electrogravitic field. With a charge of several hundred kilovolts the condenser would reach speeds of several hundred miles per hour.

speed of flight would be effected by merely altering the intensity, polarity and direction of the charge. Man now uses the sledge-hammer approach to high-altitude high-speed flight. In the still-short life-span of the turbo-jet airplane he has had to increase power in the form of brute thrust some twenty times in order to achieve just a little more than twice the speed of the original jet plane. The cost in money in reaching this point has been prodigious. The cost in highly-specialized man-hours is even greater. By his present methods man actually fights in direct combat the forces that resist his efforts. in conquering gravity he would be putting one of his most competent adversaries to work for him. Antigravitics is the method of the picklock rather than the sledge-hammer. The communications possibilities of electrogravitics, as the new science is called, confound the imagination. There are apparently in the ether an entirely new unsuspected family of electrical waves similar to electromagnetic radio waves in basic concept. Electrogravitic waves have been created and transmitted through concentric layers of the most efficient kinds of electromagnetic and electrostatic shielding without any apparent loss of power in any way. There is evidence, but not yet proof, that these waves are not limited by the speed of light. Thus the new science seems to strike at the very foundations of Einsteinium Relativity Theory. But rather than invalidating current basic concepts such as Relativity, the new knowledge of gravity will probably expand their scope, ramification and general usefulness. It is this expansion of knowledge into the unknown that more emphasizes how little we do know; how vast is the area still awaiting research and discovery. The most successful line of the electrogravitics research so far reported is that carried on by Townsend. T. Brown, an American who has been researching gravity for over thirty years. He is now conducting research projects in the U.S. and on the Continent. He postulates that there is between electricity and gravity a relationship parallel and/or similar to that which exists between electricity and magnetism. And as the coil is the usable link in the case of electromagnetics, so is the condenser that link in the case of electrogravitics. Years of successful empirical work have lent a great deal of credence to this hypothesis.

The detailed implications of man's conquest of gravity are innumerable. In road cars, trains and boats the headaches of transmission of power from the engine to wheels or propellers would simply cease to exist. Construction of bridges and big buildings would be greatly simplified by temporary induced weightlessness, etc. Other facets of work now under way indicate the possibility of close controls over the growth of plant life; new therapeutic techniques, permanent fuel-less heating units for homes and industrial establishments; new manufacturing techniques; a whole new field of chemistry. The list is endless ... and growing. In the field of international affairs, other than electrogravitics' military significance, what development of the science may do to raw material are more prone to induced weightlessness than others. These are becoming known as gravitic isotopes: Some are already quite hard to find, but others are common and, for the moment, cheap. Since these ultimately may be the vital lofting materials required in the creation of contra-gravitational fields, their value might become extremely high with equivalent rearrangement of the wealth of natural resources, balance of economic power and world geo-strategic concepts. How soon all this comes about is directly proportional to the effort that is put into it. Surprisingly, those countries normally expected to be leaders in such an advanced field are so far only fooling around. Great Britain, with her Ministry of Supply and the National Physical Laboratory, apparently has never

Author's diagram illustrating the electrogravitic field and the resulting propulsive force on a disc-shaped electrostatic condenser. The centre of the disc is of solid aluminium. The solid rimming on the sides is perspex, and in the trailing and leading edges (seen in the direction of motion) are wires separated from the aluminium core chiefly by air pockets. The wires act in a manner similar to the two plates of a simple electrical condenser and, when charged, produce a propulsive force. On reaching full charge a condenser normally loses its propulsive force, but in this configuration the air between the wires is also charged, so that in principle the charging process can be maintained as long as desired. As the disc also moves—from minus to plus—the charged air is left behind, and the condenser moves into new, uncharged air. Thus both charging process and propulsive force are continuous.

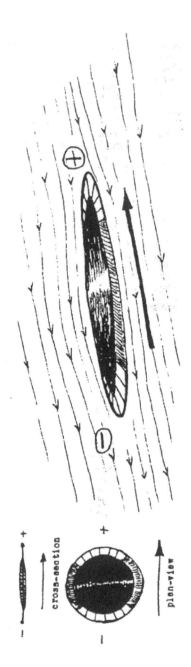

cross-section

plan-view

82

seriously considered that the attempt to overcome and control gravity was worth practical effort and is now scurrying around trying to find out what its all about. The U.S. Department of Defense has consistently considered gravity in the realm of basic theory and has so far only put token amounts of money into research on it. The French, apparently a little more open-minded about such things, have initiated a number of projects, but even these are still on pretty much of a small scale. The same is true throughout most of the world. Most of the work is of a private venture kind, and much is being done in the studies of university professors and in the traditional lofts and basements of badly undercapitalized scientists. But the word's afoot now. And both Government and private interest is growing and gathering momentum with logarithmic acceleration. The day may not be far off when man again confounds himself with his genius; then wonders why it took him so long to recognize the obvious. Of course, there is always a possibility that the unexplained 3% of UFOs, "Unidentified Flying Objects", as the U.S. Air Force calls "flying saucers", are in fact vehicles so propelled, developed already and undergoing proving flights – by whom.... U.S., Britain....or Russia? However, if this is so, it's the best kept secret since the Manhattan Project, for this reporter has spent over two years trying to chase down work on gravitics and has drawn from Government scientists and military experts the world over only the most blank of stares.

This always the way of exploration into the unknown.

115th Year   VOL. CXV
NO. 10,356

230 West 41st Street, New York 36, N. Y.
Telephone PEnnsylvania 6-4000

Conquest of Gravity Aim Of Top Scientists in U.S.

ANTI-GRAVITY RESEARCH—Dr. Charles T. Dozier, left, senior research engineer and guided missiles expert of the Convair division of General Dynamics Corp., conducting a research experiment toward control of gravity with Martin Kaplan, Convair senior electronics engineer.

Changes Far Beyond Atom Are the Prize

Revolution in Power, Air, Transit Is Seen

This is the first of a series on new pure and applied research into the mysteries of gravity and efforts to devise ways to counteract it.

By Ansel E. Talbert
Military and Aviation Editor

© 1955, New York Herald Tribune Inc.

The initial steps of an almost incredible program to solve the secret of gravity and universal gravitation are being taken today in many of America's top scientific laboratories and research centers.

A number of major, long-established companies in the United States aircraft and electronics industries also are involved in gravity research. Scientists in general bracket gravity with life itself as the greatest unsolved mystery in the uni-

IN CHARGE—George S. Trimble jr., vice-president in charge of advanced design planning of Martin Aircraft Corp., is organizing a new Research Institute for Advanced Study to push a program of theoretical research on gravitational effects.

Helicopters to Bring Cabinet

Conquest of Gravity: Aim of Top Scientists in U.S.

Written by Ansel E. Talbert, Military and Aviation Editor. N.Y.H.T.

New York Herald Tribune, Sunday, November 20, 1955, pp.1 & 36

This is the first of a series on new pure and applied research into the mysteries of gravity and efforts to devise ways to counteract it.

ANTI-GRAVITY RESEARCH - Dr. Charles T. Dozier, left, senior research engineer and guided missiles expert of the Convair Division of General Dynamics Corp., conducting a research experiment toward control of gravity with Martin Kaplan, Convair Senior electronics engineer.

(Photo inset)
IN CHARGE -George S. Trimble, Jr., vice-president in charge of advanced design planning of Martin Aircraft Corp., is organizing a new research institute for advanced study to push a program of theoretical research on gravitational effects.'. "CHANGES FAR BEYOND THE ATOM ARE THE PRIZE" (Revolution in Power, Air Transit Is Seen)

The initial steps of an almost incredible program to solve the secret of gravity and universal gravitation are being taken today in many of America's top scientific laboratories and research centers.

A number of major, long-established companies in the United States aircrafts and electronics industries also are involved in gravity research. Scientists, in general, bracket gravity with life itself as the greatest unsolved mystery in the Universe. But there are increasing numbers who feel that there must be a physical mechanism for its propagation which can be discovered and controlled.

Should this mystery be solved it would bring about a greater revolution in power, transportation and many other fields than even the discovery of atomic power. The influence of such a discovery would be of tremendous import in the field or aircraft design - where the problem of fighting gravity's effects has always been basic.

A FANTASTIC POSSIBILITY

85

One almost fantastic possibility is that if gravity can be understood scientifically and negated or neutralized in some relatively inexpensive manner; it will be possible to build aircraft, earth satellites, and even space ships that will move swiftly into outer space, without strain, beyond the pull or earth's gravity field. They would not have to wrench themselves away through the brute force of powerful rockets and through expenditure of expensive chemical fuels.

Centers where pure research on gravity is now in progress in some form include the Institute for Advanced Study at Princeton, N.J., and also at Princeton University: the University of Indiana's School of Advanced Mathematical Studies and the Purdue University Research Foundation.

A scientific group from the Massachusetts Institute of Technology, which encourages original research in pure and applied science, recently attended a seminar at the Roger Babson Gravity Research Institute of New Boston, N.H., at which Clarence Birdseye, inventor and industrialist also was present. Mr, Birdseye gave the world it's first packaged quick-frozen foods and laid the foundation for today's frozen food industry; more recently he has become interested in gravitational studies.

A proposal to establish at the University of North Carolina at Chapel Hill, N.C., an 'Institute of Pure Physics' primarily to carry on theoretical research on gravity was approved earlier this month by the University's board of trustees. This had the approval of Dr. Gordon Gray who has since retired as president of the University; Dr. Gray has been Secretary of the Army, Assistant Secretary of Defense, and special assistant to the President of the United States.

FUNDS COLLECTED

Funds to make the institute possible were collected by Agnew H. Bahnson Jr., an industrialist or Winston Salem. N.C. The new University or North Carolina administration is now deciding on the institute's scope and personnel. The directorship has been offered to Dr, Bryce DeWitt of the Radiation Laboratory at the University of California at Berkeley, Who is the author or a Roger Babson prize-winning scientific study entitled "New Directions for Research in the Theory of Gravity."

The same type or scientific disagreement which occurred in connection with the first proposals to build the hydrogen bomb and an artificial earth satellite - now under construction - is in progress over anti-gravity research. Many scientists of repute are sure that gravity can be overcome in comparatively few years if sufficient resources are

86

put behind the project. Others believe it may take a quarter of a century or more.

REFUSE TO PREDICT

Some pure physicists, while backing the general program to try to discover how gravity is propagated, refuse to make predictions of any kind.

Aircraft industry firms now participating or actively interested in gravity include the Glenn. L. Martin Co. of Baltimore, builders of the nations First giant jet-powered flying boat; Conair of San Diego, designers and builders of the giant B-36 intercontinental bomber and the world's first successful vertical take-of fighter; Bell Aircraft of Buffalo, builders or the first piloted airplane to fly faster than sound and a current jet vertical take of and landing airplane and Sikorsky division of United Aircraft, pioneer helicopter builders.

Lear, Inc., of Santa Monica, one of the world's largest builders of automatic pilots for airplanes; Clarke Electronics of Palm Springs, California, a pioneer in its field, and the Sperry Gyroscope Division or Sperry-Rand Corp. of Great Neck, L.I., which is doing important work on guided missiles and earth satellites, also have scientists investigating the gravity problem.

USE EUROPEAN EXPERT

Martin Aircraft has just put under contract two of Europe's leading theoretical authorities on gravity and electromagnetic fields – Dr. Burkhard Heim of Goettingen University, where some of the outstanding discoveries of the century in aerodynamics and physics have been made and Dr. Pascual Jordan of Hamburg University, Max Planck medal winner whose recent work called: 'Gravity and the Universe' has excited scientific circles throughout the world.

Dr Heim, now professor of theoretical physics, at Goettingen and who was a member of Germany's Bureau of Standards, during World War II, is certain that gravity can be overcome. Dr. Heim, lost his eyesight and hearing, and had both his arms blown off at the elbows, in a World War II rocket explosion. He dictates his theories and mathematical calculations to his wife.

Martin Aircraft, at the suggestion of George S Trimble, its vice president in charge of advanced design planning, is building between Washington and Baltimore a new laboratory for the Research Institute for Advanced Study... A theoretical investigation of the implications for future gravity research in the 'Unified Field Theory' of the late Dr Albert Einstein is now underway here.

87

Although financed by Martin, the Institute will have no connection with the day-to-day business of building airplanes. Its general manager is Welcome Bender. Up to now no scientist or engineer, so far as is known in the scientific circles, has produced the slightest alteration in the magnitude or direction of gravitational 'force' although many cranks and crackpots have claimed to be able to do this with 'perpetual motion machines.'

NO ACCEPTED THEORY

There is no scientific knowledge or generally accepted theory about the speed with which it travels across interplanetary space, making any two material particles or bodies - if free to move - accelerate toward each other.

But the current efforts to understand gravity and universal gravitation both at subatomic level and at the level of the Universe have the positive backing today of many of America's outstanding physicists.

These include Dr. Edward Teller of the University of California, who received prime credit for developing the hydrogen bomb; Dr. J. Robert Oppenheimer, director of the Institute for Advanced Study at Princeton; Dr. Freeman J. Dyson, theoretical physicist at the Institute, and Dr. John A. Wheeler, professor of physics at Princeton University, who made important contributions to America's first nuclear fission project.

PURE RESEARCH VIEW

It must be stressed that scientists in this group approach the problem only from the standpoint of pure research, They refuse to predict exactly in what directions the search will lead or whether it will be successful beyond broadening human knowledge generally.

Other top-ranking scientific minds being brought to bear today on the gravity problem are those of Dr. Vaclav Hlavaty, of the University of Indiana, who served with Dr. Einstein on the faculty of Charles University in Prague and later taught advanced mathematics at the Sorbonne in Paris; and of Dr. Stanley Deser and Dr. Richard Arnowitt of the Princeton Institute for Advanced Study.

Dr. Hlavaty believes that gravity simply is one aspect of electromagnetism, the basis of all cosmic forces, and eventually may be controlled like light and radio waves.

HOPE TO FIND KEY

Dr. Deser and Dr. Arnowitt are of the opinion that very recently discovered nuclear and sub-nuclear particles of high energy, which are difficult to explain by any present-day theory, may prove to be the key that eventually unlocks the mystery. It is their suggestion that the new particles may prove to be basic gravitational energy which is being converted continually and automatically in an expanding Universe 'directly into the most useful nuclear and electromagnetic forms.' In a recent scientific paper they point out: 'One of the most hopeful aspects of the problem is that until recently gravitation could be observed but not experimented on in any controlled fashion, while now with the advent in the past two years of the new high-energy accelerators (the Cosmotron and the even more recent Berkeley Bevatron) the new particles which have been linked with the gravitational field can be examined and worked with at will.'

An important job of encouraging both pure and applied gravity research in the United States through annual prizes and seminars, as well as the summarizing of new research for engineers and scientists in industry looking forward to a real "hardware solution" to the gravity problem, is being performed by the Gravity Research Foundation of New Boston N.H. This was founded and funded by Dr Roger Babson, economist, who is an alumnus of MIT and a lifelong student of the works of Sir Isaac Newton, discoverer of gravity. Its president is Dr George Rideout of Boston.

(Photo Insets):
BLACKBOARD MATH
Dr. Vaclav Hlavaty, of the University of Indiana's graduate Institute of Advanced Mathematics, who has stimulated research on gravity control, working on a problem.

ANTI-GRAVITY AND AVIATION –George S. Trimble Jr., vice-president in charge of advanced design planning of Martin Aircraft Corp., discussing the application of anti-gravitational research to aviation with two Martin scientists, J.D. Pierson, centre, and William B Yates.

End Article One

BLACKBOARD MATH—Dr. Vaclav Hlavaty, of the University of Indiana's Graduate Institute of Advanced Mathematics, who has stimulated research on gravity control, working on a problem.

Gravity

(Continued from preceding)

nology, which encourages dis-inal research in pure and ap-

leading theoretical authorities on gravity and electromagnetic fields Dr. Burkhard Heim, of Goettingen University, where some of the outstanding dis-coveries of the century in aero-dynamics and physics have been

ANTI-GRAVITATION AND AVIATION—George S. Trimble jr., vice-president in charge of advanced design planning of Martin Aircraft Corp., left, discussing the application of anti-gravitational research to aviation with two Martin scientists, J. D. Pierson, center, and William B. Yates.

gineer—so far as is known in the University of Indiana, who industry looking forward to a scientific circles—has produced served with Dr. Einstein on the real "hardware solution" to the the slightest alteration in the faculty of Charles University in gravity problem is being per- magnitude or direction of gravi- Prague and later taught ad- formed by the Gravity Research

Space-Ship Marvel Seen if Gravity Is Outwitted

Ansel E. Talbert. Military and Aviation Editor, N.Y.H.T
New York Herald-Tribune: Monday, November 21, 1955. pp. 1&6

This is the second of a series on new pure and applied research into the mysteries of gravity and efforts to devise ways to counteract it.

(Photo Inset):
FLYING SAUCER OF THE FUTURE? - A reproduction of an oil painting by Eugene M. Gluhareff, president of Gluhareff Helicopter & Airplane Corp of Manhattan Beach, California, showing "saucer-shaped" aircraft or space ship for exploring far beyond the earth's atmosphere and gravity field. Mr. Gluhareff portrays it operating at moderate speed' over the New York - New England area and and notes that in the painting a propulsive blast of the electron beams from the rear of the saucer is visible, giving the saucer a translational force.

chance of snow flurry; clearing in afternoon; fair, cool tonight.

Tomorrow: Increasing cloudiness, milder, chance of rain at night.

Temperatures Yesterday: Max. 27.3; Min. 20.7
Today's Probable Range: Max. 47, Min. 29
Humidity at 2 p. m. Yesterday: 38%
Expected Humidity This Afternoon: 66-70%
Reports, Maps, Sec. 3, Page 4

Herald

FLYING SAUCER OF THE FUTURE? A reproduction of an oil painting by Eugene M. Gluhareff, president of Gluhareff Helicopter & Airplane Corp. of Manhattan Beach, Calif., showing a "saucer-shaped" aircraft or space ship for exploring far beyond the earth's atmosphere and gravity field. Mr. Gluhareff portrays it operating "at moderate speed" over the New York-New England area and notes that in the painting "a propulsive blast of the electron beams from the rear of the saucer is visible, giving the saucer a translational force."

Space-Ship Marvel Seen If Gravity Is Outwitted

Speeds of Thousands of Miles An Hour Without a Jolt Held Likely

This is the second of a series on new pure and applied research into the mysteries of gravity and efforts to devise

dent of the United States "for the greatest achievement in aviation in America" through developing a lightweight auto-

Republicans Split Over Labor Issue

Dispute May Go To Eisenhower

SPEEDS OF THOUSANDS OF MILES AN HOUR WITHOUT A JOLT HELD LIKELY

Scientists today regard the earth as a giant magnet. Many in America's aircraft and electronics' industries are excited over the possibility of using its magnetic and gravitational fields as a medium of support for amazing 'flying vehicles' which will not depend on the air for lift.

Space ships capable of accelerating in a few seconds, at speeds many thousands of miles an hour and making sudden changes of course at these speeds without subjecting their passengers to the so-called G forces, caused by gravity's pull also are envisioned. These concepts are part of a the program to solve the secret of gravity and universal gravitation already in progress in many top scientific laboratories and long-established industrial firms of the nation.

NUCLEAR RESEARCH AIDS

Although scientists still know little about gravity and its exact relationship to electromagnetism, recent nuclear research and experiments with 'high energy machines' such as the Brookhaven Cosmotron, are providing a flood of new evidence believed to have a bearing on this.

William P. Lear, inventor and chairman of the board of Lear. Inc., one of the nation's largest electronics firms specializing in aviation, for months has been going over new developments and theories relating to gravity with his chief scientists and engineers.

Mr. Lear in 1950 received the Collier Trophy from the President of the United States for the greatest achievement in aviation in America' through developing a lightweight, automatic pilot and approach control system for jet fighter planes. He is convinced that it will be possible to create artificial "electro-gravitational fields" whose polarity can be controlled to cancel out gravity.

He told this correspondent: "All the (mass) materials and human beings within these fields will be part of them. They will be adjustable so as to increase or decrease the weight of any object in its surroundings. They won't be affected by the earth's gravity or that of any celestial body. ... 'This means that if any person was in an anti-gravitational airplane or space ship that carried along its own gravitational field no matter how fast you accelerated or changed course – your body wouldn't any more feel it than it now feels the speed of the earth.'

Scientists and laymen for centuries have been familiar with the phenomena that 'like' poles or two magnets – the north and the north poles for example – repel each other while 'unlike' poles exert an attraction. In ancient times 'lodestones' possessing natural magnetism were thought to possess magical powers.

FARADAY'S DISCOVERIES

But the nineteenth century discoveries of England's great scientist, Michael Faraday, paved the way for construction of artificial 'electro-magnets' – in which magnetism is produced by means or electric currents. They retain it only so long as the current is flowing. An electromagnet can be made by winding around a soft iron 'core' a coil of insulated wire carrying electric current. Its strength depends primarily on the number of turns in the coil rather than the strength of the current.

Even today, America's rapidly expanding electronics industry is constantly finding new uses for electromagnets. For example, Jack Fletcher, a young electronics and aeronautical engineer of Covina, Calif., has just built a 'Twenty-First Century Home' containing an electronic stove functioning by magnetic repulsion.

PAN FLOATS IN AIR

In it seven coils of wire on laminated iron cores are contained inside a plywood cabinet or blond mahogany. The magnetic field from these coils induces 'eddy currents' in an aluminum cooking pan nineteen inches in diameter, which interact and lift the pan into space like a miniature 'flying saucer.'

The cooking pan floats about two inches in the air above the stove in a stabilized condition; 'eddy currents' generate the heat that warms it while the stove top itself remains cold. The aluminum pan will hold additional pots and it can be used as a griddle. It is, of course, a variation of several other more familiar magnetic repulsion gadgets including the 'mysterious floating metal ball' or science hall exhibits. No type of electromagnet known to science or industry would have any application to the building of a real aircraft or 'flying vehicle' . But one of America's most brilliant young experimental designers, Eugene M. Gluhareff, president of Gluhareff Helicopter and Airplane Corp. of Manhattan Beach. Calif. has made several theoretical design studies of round or saucer-shaped 'vehicles, for travel into outer space, having atomic generators as their basic 'engines.'

SON OF COPTER DESIGNER

Mr. Gluhareff is the son of Michael E, Gluhareff, chief designer for Dr. Igor I. Sikorsky, helicopter and multi-engined aircraft pioneer . Dr. Sikorsky and the elder Mr. Gluhareff, who has won the Alexander Klemin award, one of aviation's highest honors, are themselves deeply concerned in the problem of overcoming gravitation.

The younger Mr. Gluhareff already has been responsible for several successful advanced designs of less amazing 'terrestrial' aircraft. He envisions the electric power obtained from the atomic generators operating electronic reactors - that is, obtaining propulsion by the acceleration of electrons to a very high velocity and expelling them into space in the same manner that hot gases are expelled from jet engines. Such an arrangement would not pollute the atmosphere with radioactive vapors.

COULD CONTROL ACCELERATION

Because of its 'long-lasting fuel', an atomic-electronic flying disk would be able to control its acceleration to any speed desired and there would be no need for being 'shot into space' according to Mr. Gluhareff. Radial electronic beams around the saucer's rim would be operating constantly and would sustain flight by 'acting against gravity.'

Mr. Gluhareff thinks that control can be achieved by a slight differentiation of the deflection of electronic beams in either direction; the beams would act in the same way as an orthodox plane's ailerons and elevator.

GRAVITATIONAL CHANGES

Mr. Gluhareff agrees with Dr. Pascual Jordan of Hamburg University, one of Europe's outstanding authorities on gravitation who proved many parts of the 'Quantum Theory' of Dr. Max Planck, that it will be possible to induce substantial changes in the Gravitational fields of rotating masses through electromagnetic research. Dr. Jordan has just signed a contract to do research for Martin Aircraft Corp. of Baltimore.

Norman V. Peterson guided missiles engineer of the Sperry-Gyroscope Division of Sperry-Rand Corp. of Great Neck. L.I., who as president of the American Astronautical Society attended the recent 'earth satellite' meeting in Copenhagen, corroborates the theory that 'nuclear powered - or solar powered - ion electron beam reactors - will give impetus to the conquest of space.'

(Photo Inset):

FLOATING COOKING PAN - The 'electronic stove' functioning by magnetic repulsion built by Jack Fletcher, a young engineer of West Covina, Calif., The aluminum cooking pan, nineteen inches in diameter, floats two Inches above the cabinet like a miniature 'flying saucer'. It is completely stable while 'hovering' and can be used as a griddle or as a holder for additional pots and pans. 'Eddy currents' from a magnetic field created by an electromagnet inside the cabinet have warmed the pan - although the stove top remains completely cold.

End Article Two

NEW YORK HERALD TRIBUNE, MONDAY, NOVEMB

Gravity

(Continued from page one)

was in an anti-gravitational plane or space ship that car-i along its own gravitational d—no matter how fast you elerated or changed course—r body wouldn't any more it than it now feels the 'd of the earth."

cientists and laymen for cen-es have been familiar with phenomena that "like" poles wo magnets—the north and h poles for example—repel 1 other while "unlike" poles t an attraction. In ancient 's "lodestones" possessing iral magnetism were thought ossess magical powers.

Faraday's Discoveries

it the nineteenth century overies of England's great tist, Michael Faraday, paved way for construction of ar-al "electro-magnets" — in h magnetism is produced by is of electric currents. They n it only so long as the cur-is flowing. An electro-net can be made by wind-around a soft iron "core" 1 of insulated wire carrying rical current. Its strength nds primarily on the num-of turns in the coil rather the strength of the current en today America's rapidly ading electronics industry nstantly finding new uses electro-magnets. For ex-, Jack Fletcher, a young onics and aeronautical en-r of West Covina. Calif., just built a "Twenty-first iry Home" containing an onic stove functioning by

FLOATING COOKING PAN—The "electronic stove" functioning by magnetic repulsion built by Jack Fletcher, a young engineer of West Covina, Calif. The aluminum cooking pan, nineteen inches in diameter, floats two inches above the cabinet like a miniature "flying saucer." It is completely stable while "hovering" and can be used as a griddle or as a holder for additional pots and pans. "Eddy currents" from a magnetic field created by an electromagnet inside the cabinet have warmed the pan—although the stove top remains completely cold.

New Air Dream–Planes Flying Outside Gravity

Ansel E. Talbert, Military and Aviation Editor, N. Y .H. T.
New York Herald Tribune: Tuesday, November 22, 1955, pp. 1 & 10

This is the third in a series of three articles on new pure and applied research into the mysteries or gravity and the efforts to devise ways to overcome it.

Able to Go 'Where We Want'

New Air Dream—Planes Flying Outside Gravity

This is the third of a series of three articles on new pure and applied research into the mysteries of gravity and the efforts to devise ways to overcome it.

By Ansel E. Talbert
Military and Aviation Editor

The current interest in America's aircraft and electronics industries in finding out whether gravity can be controlled or "cancelled out" is not confined to imaginative young graduates of engineering and scientific schools.

Some of the two industries'

INDEX

most experienced and highly regarded leaders today are engaged directly or deeply interested in theoretical research relating to gravity and universal gravitation. Their basic aim is eventually to build "hardware" in the shape of planes, earth satellites and space ships "which can go where we want and do what we want without interference from gravity's mysterious trans-spatial pull."

Bell Is Optimistic

Lawrence D. Bell, whose company in Buffalo built the first piloted aircraft in history to fly faster than sound is certain that

(Photo Inset):
LAWRENCE D. BELL, founder and president of Bell Aircraft Corp. of Buffalo, using a Japanese ivory ball to illustrate his view that humans before long will operate planes outside the earth's atmosphere, then outside the gravity field of the earth. The pilots with him, three top test pilots or the Air Force, are, left, Lt. Col. Frank J. Everest;

97

centre, in light suit, Maj. Charles Yeager, and, in uniform next to Mr. Bell, Maj. Arthur Murray.

ABLE TO GO WHERE WE WANT

The current interest in America's aircraft and electronics industries in finding whether gravity can be controlled or "canceled-out" is not confined to imaginative young graduates of engineering and scientific schools.

Some of the two industries' most experienced and highly regarded leaders today are engaged directly or deeply interested in theoretical research relating to gravity and universal gravitation. Their basic aim is eventually to build 'hardware' in the shape of planes, earth satellites, and space ships, which can go where we want and do what we want without interference from gravity's mysterious trans-spatial pull.

NEW YORK HERALD TRIBUNE, TUESDAY, NOVEMBER 22, 1955

Lawrence D. Bell, founder and president of Bell Aircraft Corp., of Buffalo, using a Japanese ivory ball to illustrate his view that humans before long will operate planes outside the earth's atmosphere, then outside the gravity field of the earth. The pilots with him, three top test pilots of the Air Force, are, left, Lt. Col. Frank J. Everest, centre, in light suit, Maj. Charles Yeager, and, in uniform next to Mr. Bell, Maj. Arthur Murray.

Gravity

98

BELL IS OPTIMISTIC

Lawrence D. Bell, whose company in Buffalo built the first piloted aircraft in history to fly faster than sound, is certain that practical results will come out of current gravity research. He told this correspondent: 'Aviation as we know it is on the threshold of amazing new concepts. The United States aircraft industry already is working with nuclear fuels and equipment to cancel out gravity instead of fighting it. The Wright Brothers proved that man does not have to be earth-bound. Our next step will be to prove that we can operate outside the earth's atmosphere and the third will be to operate outside the gravity of the earth.'

OPTIMISM SHARED

Mr. Bell's company, during the last few days made the first powered flights with its new Bell X-2 rocket plane designed to penetrate deep into the thermal or heat barrier encountered due to atmospheric fiction at a speed above 2,000 miles per hour. It also is testing a revolutionary new jet vertical-rising-and-landing 'magic carpet' airplane.

Grover Loening, who was the first graduate in aeronautics in an American University and the first engineer hired by the Wright Brothers, holds similar views.

Over a period of forty years, Mr. Loening has had a distinguished career as an aircraft designer and builder recently was decorated by the United States Air Force for his work as a special scientific consultant.

'I firmly believe that before long man will acquire the ability to build an electromagnetic contra-gravity mechanism that works." he says. 'Much the same line of reasoning that enabled scientists to split up atomic structures also will enable them to learn the nature of gravitational attraction and ways to counter it.'

Right now there is considerable difference of opinion among those working to discover the secret of gravity and universal gravitation as to exactly how long the project will take. George S Trimble, a brilliant young scientist who is head of the new advanced design division of Martin Aircraft in Baltimore and a member of the sub-committee on high-speed aerodynamics of the National Advisory Committee for Aeronautics, believes that it could be done relatively quickly if sufficient resources and momentum were put behind the program. "I think we could do the job in about the time that it actually required to build the first atom bomb if enough trained scientific brain- power simultaneously began thinking about and working towards a solution," he said. "Actually, the biggest deterrent to scientific progress is a refusal of some people, including scientists, to believe that things which seem amazing can really happen. "I know that if Washington decides that it is vital to our national survival to go where we want and do what we without having to worry about gravity, we'd find the answer rapidly."

SIKORSKY CAUTIOUS

Dr. Igor I. Sikorsky, one of the world's outstanding airplane and helicopter designers, is somewhat more conservative but equally interested. He believes that within twenty-five years man will be flying beyond the earth's atmosphere, but he calls gravity, 'real, tangible, and formidable.' It is his considered scientific observation that there must be some physical carrier for this immense trans-spatial force.

Dr. Sikorsky notes that light and electricity, once equally mysterious, now have become 'loyal, obedient servants of man, appearing and disappearing at his command and performing at his will a countless variety of services.' But in the case of gravitation he says the more scientists attempt to visualize the unknown agent which transmits it, 'the more we recognize we are facing a deep and real mystery.

The situation calls for intensive scientific research. Dr. Sikorsky believes. Up to now all gravity research in the United States bas been financed out of the private funds of individuals or corporations. Leaders of the nation's armed forces have been briefed by various scientists about the theoretical chances of conquering gravitation but so far their attitude is call us when you get some hardware that works.

Dudley Clarke, president of Clarke Electronics laboratories of Palm Springs, Calif., one of the nation's oldest firms dedicated electronic research and experimentation, is one scientist in the hardware stage of building something that he believes will prove gravity can be put to useful purposes.

Mr. Clarke's company has just caused astir in the electronics industry by developing pressure-sensitive resistors having unusual characteristics for parachute and other aviation use, according to "Teletech and Electronic Industries" magazine of 480 Lexington Ave.

Mr. Clarke who years ago worked under Dr. Charles Steinmetz, General Electric Company's electrical and mathematical 'wizard' of the 1930's, is sure that this successful harnessing of gravitation will take place sooner than some of these 'ivy tower' scientists believe.

Like Sir Frank Whittle, Britain's jet pioneer, who was informed in 1935 by the British Air Ministry that it could see no practical use for

100

his jet aircraft engine, Mr. Clarke has a particularly cherished letter. It was written about the same time by the commanding general at Wright Field giving a similar analysis of a jet design proposal by Mr. Clarke. Mr. Clarke notes that the force of gravity is powerful enough to generate many thousand times more electricity than now is generated at Niagara Falls and every other water-power centre in the world - if it can be harnessed. This impending event, he maintains, will make possible the manufacture of anti-gravity 'power packages' which can be bought for a few hundred dollars. These would provide all the heat and power needed by one family for an indefinite period.

Dr. W.R.G. Baker, vice-president and general manager of General Electric, electronics division, points out that scientists working in many fields actually are beginning to explore the universe, learning new things about the makeup of 'outer space' and formulating new concepts. He says: 'Today we in electronics are deeply interested in what lies beyond the earth's atmosphere and its gravity field. For there we may find the electronics world of what now. Such questions usually have been reserved for the realm of physics and astronomy. But through entirely new applications in radar for example science already is able to measure some of the properties of the world beyond. Warm bodies radiate microwaves, and by recording noise signals', we are learning about invisible celestial forces we did not know existed.'

Dr. Arthur L. Klein, professor of aeronautics at the California Institute of Technology, is certain that 'if extra-terrestrial flight is to be achieved, something will be required to replace chemical fuels.'

Dr. Hermann Oberth, Germany's greatest rocket pioneer, who is now working on guided missiles for the United States Army, calculates that 40,000 tons of liquid propellants will be required to lift a payload of only two tons beyond the earth's gravitation. Regarding this chemical fuel problem Dr. Klein says, 'there are no other serious obstacles.'

Many thoughtful theoretical scientists and practical engineers see a space vehicle de-gravitized to a neutral weight and following an electronically-controlled route, charted by radar as the ultimate answer."

End Article Three

Project Winterhaven – For Joint Services R&D Contract

By T. Townsend Brown

PURPOSES

For the last several years, accumulating evidence along both theoretical and experimental lines has tended to confirm the suspicion that a fundamental interlocking relationship exists between the electrodynamic field and the gravitational field.

It is the purpose of Project WINTERHAVEN to compile and study this evidence and to perform certain critical or definitive experiments which will serve to confirm or deny the relationship. If the results confirm the evidence, it is the further purpose of Project WINTERHAVEN to examine the physical nature of the basic electrogravitic couple" and to foresee and develop possible long-range practical applications.

The proposed experiments are to be limited at first to force measurements and wave propagation. They are to be expanded, depending to include applications in propulsion or motive power, communications and remote control, with emphasis on military applications of recognized priority.

The story of the falling apple, which led to Sir Isaac Newton's law of gravitation, is familiar to nearly everyone. It is the usual starting point in any resume about gravitation.

Newton's law was the first mathematical expression of a strange and mysterious force – a force which has continued to remain a mystery for over two hundred years.

During this period, few scientists have emerged to offer a solution – so great, as a matter of fact, has been the enigma. In the dusty unpublished notes of Sir Oliver Heaviside, written in the latter part of the nineteenth century, a remarkably adequate theory of gravitation was proposed. It was the first theory, so far as is known to link the electrodynamic field to the gravitational field.

In 1905, Einstein published the Special Theory of Relativity and this was soon followed by the General Theory, describing gravitation in quite different terms but again implying a similarity and possible relationship with the electrodynamic field. Subsequently, in the Unified Field Theories, Einstein has attempted to work out the

102

mathematical basis for such a correlation, but so far has been unable to offer any specific experiment or observation (as in the case of Relativity) by which such a suspected relationship can be proved.

Compelled by a deep interest in the subject, Townsend Brown (as an 18-year old student at the California Institute of Technology and later at Denison University) performed crude but apparently significant experiments with electric capacitors, using plates and dielectrics of various mass. The impetus for such an investigation was provided by the academic controversy which Relativity aroused in the early twenties. Brown developed the thesis that, due to the similar or equivalent nature of the electric and gravitational fields, a reciprocal influence could be expected which, if constrained, would give rise to physical forces detectable under certain circumstances.

These early studies and the experimental results were called to the attention of Dr. Paul Alfred Biefeld (a colleague of Albert Einstein in Germany. See appendix "Who's Who", then professor of astronomy at Denison University and director of Swazey Observatory). Dr. Biefeld continued his interest and active support of the experiments for many years and, prior to his untimely death in 1936, subscribed by affidavit that the observed effects in his opinion did represent "an influence of the electrostatic field upon the gravitational field". This strange new effect first indicated by the results of these experiments with electric capacitors, has since been named the Biefeld-Brown effect, but due to the uncompleted experiments and inconclusive results, publication has been withheld. In recent years, as additional data of a confirming nature became available, the research has been associated with government research projects of a highly classified status, and publication has been precluded.

Townsend Brown continued to conduct studies of this basic effect with particular attention to increasing the ponderomotive forces revealed in massive dielectric materials, especially, as it became apparent, in those materials with high specific inductive capacity or dielectric constant (K). Various obstacles were met and were only partly overcome. There remained the problem of supplying the required high potentials and developing suitable dielectric materials capable of withstanding such potentials.

Due largely to the limitation of the dielectric constant (K) of materials available in those days, the forces obtained in the early stages of the research were never very large. Hence the effect remained for many years in the category of a "scientific curiosity." It appeared impossible to increase the "K" to a value sufficient to produce consistently measurable or mechanically useful forces.

103

Within the last few years, however, due to the demands of radar and television instrumentation, new dielectric materials have been developed. The available values of K have progressively increased from 6 to 100, from 6,000 to 30,000 and beyond. Dielectrics with K of 6,000 are now available commercially, increasing by a factor of one thousand the magnitude of the ponderomotive forces theoretically obtainable. This should be sufficient, if the theory holds, to produce mechanical forces large enough to be accurately measured and also to be useful. In short, it now appears that materials are available at last which are necessary to conduct experiments which will be conclusive in proving or disproving the hypothesis that "a gravitational field can be effectively controlled by manipulating the space-energy relationships of the ambient electrostatic field".

RESEARCH ON THE CONTROL OF GRAVITATION

In further confirmation of the existing hypothesis, experimental demonstrations actually completed in July 1950, together with subsequent confirmations with improved materials tend to indicate that a new motive force, useful as a prime mover, has in reality been discovered. While the first experiments with new dielectric materials of higher K indicated the presence of a noteworthy force, the tests were mainly qualitative and imperfect because of other factors, and the ultimate potential in terms of thrust still remains highly theoretical. The behavior of the new motive force nevertheless does appear to be in agreement with the hypothesis that there is an interaction between the electrical field and the gravitational field and that this interaction may be electrically controlled. Discovery of what may turn out to be the long-sought "electrogravitic couple" should lead to the development of an entirely new form of prime mover, a form of electric motor utilizing electrical and gravitational fields in combination – rather than electric and magnetic fields (as in all other forms of motors in use at the present time). It is interesting to note that virtually all of the electric industry today is based on the electro-magnetic inter-relationship in one form or another, dating back to the historic research of Faraday and Maxwell. These original formulations have been changed but little during the growth and development of the electrical age.

It is believed by the sponsors of Project WINTERHAVEN that the technical development of the electrogravitic reaction would usher in anew age of speed and power and of revolutionary new methods of transportation and communication. Theoretical considerations would predict that, because of the privilege of sustained acceleration, top limits of speed may be raised far beyond those of jet propulsion or rocket drive, " with possibilities eventually of approaching the speed of light in "free space". The motor which may be forthcoming will be essentially soundless, vibration less and heatless.

As a means of propulsion in flight, its potentialities already appear to have been demonstrated in model disc-shaped airfoils, a form to which it is ideally adapted. These model airfoils develop a linear thrust like a rocket and may be headed in any direction.

The discs contain no moving parts and do not necessarily rotate while in flight. In atmospheric air they emit a bluish-red electric coronal glow and a faint hissing sound.

Rocket-type electrogravitic reactor motors may prove to be highly efficient. Theoretically, internal resistance losses are almost negligible and speeds can be enormous. The thrust is controllable by the voltage applied, and a reversal of electric polarity may even serve as a brake (or if maintained, reverse the direction of flight).

A tentative theory of the electrogravitic motor has been fairly well worked out and seems to be substantiated in all tests to date. However, there are certain variable factors which are not completely understood. For example, there are tidal effects apparently caused by the Sun and Moon which influence to a small extent the power developed. There are anomalous sidereal effects which seem to be related to the passage of the Earth through diffuse clouds of cosmic dust or electrified particles ejected from the Sun. There is no assurance that large-scale experiments might not reveal additional unknowns, and it is felt that only by continued research and successively more advanced steps can the ultimate development be realized.

RESULTS OF RESEARCH TO DATE

The Biefeld-Brown Effect was first observed in the movement of electrically-charged massive pendulums. It was subsequently observed in the movement of electrical condensers of various mass which were similarly suspended and then charged. Mechanical forces, proportional to the mass of the charged elements, were revealed which tended to move the condensers bodily, causing them to behave as if they were "falling" in the "gravitational" sense. These early results were surprising for the reason that they failed to reveal a directional effect with respect to the gravitational field of Earth, but showed only a dependence upon the mass (m) of the electrified bodies.

In the years since the Biefeld-Brown Effect was first observed, other data have indicated this relative independence from the field of the Earth, and now a satisfactory explanation has gradually evolved which removes the apparent paradox. The result has been more fortunate than unfortunate – from an ultimate practical standpoint -for it has provided a theory for a gravitational drive virtually independent of the gravitational field of the Earth. Hence, it would follow that the acceleration and control of electrogravitic space-craft would be relatively unaffected upon leaving the gravitational influence of the Earth.

Several forms of electrostatically-powered motors have been designed which have seemed to indicate various degrees of gravitational characteristics. However, even the best efforts have been crude and the results complicated and difficult to analyze.

In general, two types of motors have been built: those with internal dielectric and those with external dielectric. The Townsend Brown Differential Electrometer, an automatic recording device which has been operating satisfactorily for many years, is an example of the former type. The various small models of boat motors which have been constructed are also of this type. The disc airfoils are of the second type, and these show rather surprising laboratory performance, but are extremely complex theoretically.

Captive disc airfoils 2 feet in diameter, operating at 50 KV have been found to develop a speed of approximately 17 feet per second in full atmospheric pressure. The speed appears to be at least proportional to the voltage applied and probably to some as-yet unknown exponent of the voltage.

Based on rough extrapolations from performance charts of laboratory models, the estimated speed of larger non- captive flying discs operating at 5000 KV may be 1150 miles per hour even with atmospheric resistance. It seems not unreasonable to believe that, with voltages and equipment now available, speeds in excess of 1800 miles per hour may be reached by proportionately larger discs operating at the same voltage in the upper atmosphere.

ELECTROGRAVITATIONAL COMMUNICATION SYSTEM
(Electrogravitic induction between systems of capacitors involving propagation and reception of gravitational waves)

Project started at Pearl Harbor in 1950. Theoretical background examined and preliminary demonstrations witnessed by Electronics

Officer and Chief Electronics Engineer at Pearl Harbor Navy Yard Receiver already constructed - detects cosmic noise which, according to supporting evidence, appears to emanate from that portion of the sky near the constellation Hercules (16h RA, 400 N Decl.). Transmitter designed and now partly completed. Radiation is more penetrating than radio (has been observed to pass readily through steel shielding and more than 15 feet of concrete).

In 1952 a short-range transmitting and receiving system was completed and demonstrated in Los Angeles. Transmission of an actual message was obtained between two rooms – a distance of approximately 35 feet.

Transmission was easily obtained through what was believed to be adequate electromagnetic shielding, but this test must bear repeating under more rigorous control. See definitive experiments (Group B) hereinafter proposed.

DEFINITIVE EXPERIMENTS

Group A - FIELD RELATIONSHIPS

Purpose:

The tentative theory implies that the basic relationship between the electrodynamic field and the gravitational field is revealed "during the process of charging or discharging electric capacitors."

Proposal:

A basic experiment is proposed in which two or larger high-voltage capacitors are associated spatially with a standard geophysical gravimeter. Careful observations are made of the gravitational anomalies induced in the region which accompany the change in electrical state. Studies are proposed of the effects of varying total capacitance, rate-of-change of electric charge, mass of dielectric materials, specific inductive capacity (K) of such materials and whether the spatial effects are vector or scalar. These investigations shall be directed toward the derivation of a satisfactory mathematical equation including all of the above factors.

This work is to be augmented by basic studies on variations in Earth charge (believed to be caused by natural electrogravitic induction) to be carried on by Stanford Research Institute in cooperation with the Division of Statistical Analysis of the Bureau of Standards.

Group B - WAVE PROPAGATION

Purpose:

Preliminary experiments have indicated the existence of an inductive inter-action between two independent shielded capacitors. In these experiments, a discharging capacitor induces a voltage in an adjacent capacitor and the effect appears to penetrate electromagnetic shielding. Theoretically, this effect of one capacitor upon another appears to be of electrogravitic nature and constitutes evidence of a new type of wave propagation which may eventually be utilized in a completely new method of wireless communication.

Proposal:

It is proposed that progressively larger-scale and longer-range transmissions be conducted. Beginning with untuned systems, laboratory tests are proposed to explore the basic electrogravitic relationships between simple systems of capacitors. Then, progressing to tuned systems, and pulsed (radar) applications, large-scale out-of-door demonstrations are suggested. Such demonstrations shall be conducted between suitably protected transmitting and receiving rooms (preferably underground) which are thoroughly shielded against electromagnetic (radio) radiation. Appropriate studies of wave attenuation due to transmission through sea water and large masses of earth may then also be undertaken.

This work is augmented by the basic studies on massive high-K dielectrics proposed for the University of Chicago. Calibration of receivers for natural cosmic noise or terrestrial variables is to be done at Stanford Research Institute, Menlo Park, California.

Group C - PONDEROMOTIVE FORCES IN SOLID DIELECTRICS

Purpose:

Investigations started in 1923 to ascertain "reasons for the movement of charged capacitors" point to the existence of a hitherto unrecognized ponderomotive force in all dielectrics under changing electric strain. This force appears to be a function of the specific inductive capacitance (K) and the density or mass (m) of the dielectric material, as well as voltage factors. Recent availability of the massive barium titanate high-K dielectrics give promise of developing these forces to the point where they may become of practical importance in specific propulsion applications.

Proposal:

Beginning with a careful mathematical analysis of the Townsend Brown Differential Electrometer (an instrument developed at the University of Pennsylvania and at the Naval Research Laboratory and which has been in almost continuous operation for over 20 years), studies are proposed of the forces developed in mica, glass, marble, phenolics and dielectrics in general and then, in particular, the newer barium titanate ceramic dielectrics. It is proposed that laboratory scale models of both rotary and linear "motors" be constructed and subjected to exhaustive performance tests. After suitable preliminary engineering development, it is suggested that a 500 lb. motor be constructed to propel a model ship, as a practical demonstration of one of the possibilities of the electrogravitic drive.

This work is to be augmented by basic studies of the original Biefeld-Brown experiments, conducted under carefully shielded and controlled conditions in vacuum or under oil. It is proposed that these supporting studies be carried on as pure research projects at the University of Chicago.

The space-couple experiments, including a repetition of the classic Trouton-Noble experiment but using high-K, dielectrics, are to be performed at The Franklin Institute in Philadelphia under Dr. C. T. Chase (For the participation of The Franklin Institute, see Appendix). Low-temperature experiments (using the liquid-helium cryostat) are likewise proposed for The Franklin Institute. These studies, under the personal supervision of Dr. W. F. G. Swann, are to be so designed as to provide answers to certain questions relative to the fundamental nature of gravitation. They are to embrace such subjects as the "Anomalous Mass of the Electron in Metals" and the "Behavior of Super-cooled Massive Dielectrics."

A special library project, housed at the Franklin Institute and supervised by Dr. Swann, is to serve as a clearing house and repository for information on the subject of field theories and gravitation. Whenever indicated, consultations on mathematical considerations, field theories and implications of Relativity are to be held with the Institute for Advance Studies at Princeton.

Group D -REACTIVE FORCES IN FLUID DIELECTRICS

Purpose:

Studies of boundary forces (where electrodes are in contact with fluid dielectrics) reveal the existence of a "complex" of inter-acting forces, some of which are purely electrostatic, some electromagnetic and some which could be electrogravitic. The tentative theory requires these electrogravitic forces to be present wherever a mass of dielectric material is charged and moving, and to increase in almost direct proportion to the volume of the fluid which is charged and moved. Hence it is, in a sense, the juxtaposition of the elements of the static form of capacitor described in Group C experiments, and provides what may be described as an electrokinetic propulsive system, with possible applications to high-speed aircraft and spacecraft.

Proposal:

It is proposed that electrically-charged circular airfoils be mathematically analyzed and improved. Starting with 2 ft. discs at 50 KV, the steps of the development should include 4 ft. discs at 150 KV and a final 10 ft. disc at 500 KV. Careful measurements are to be made of both static and dynamic thrust. Studies are also proposed wherein the discs are adapted for vertical lift (levitation) as well as for horizontal thrust and this feature may be incorporated in the design of the 10 ft. demonstration model.

It is proposed that studies likewise be made of various methods for obtaining the required high voltages, and these studies should include the development and evaluation of the capacitor voltage multiplier and the "flame-jet" electrostatic generator (to provide up to 15 million volts).

This work is to be augmented by the pure research projects, which are proposed for the University of Chicago, to answer certain questions as to relative efficiency of propulsion of discs in air at reduced pressure or in vacuum and at various voltages.

IMMEDIATE USES IF EXPERIMENTS PROVE TO BE POSITIVE

Confirmation of the existence of the electrogravitic couple may provide basic facts and figures which could lay the groundwork for major advances in propulsion and communication. It would initiate

110

changes in existing concepts of the theory of Relativity and the physical nature of gravitation, and certainly provide a basis for utilizing, in a practical way, hitherto unrecognized principles. It would start a major revolution in the science of physics, with profound repercussions in astronomy, chemistry and biology. In its timeliness and provocative influence, it may become a "shot heard round the world."

Propulsion:

Mankind has shown a persistent aptitude to devise means for traveling at ever-increasing rates of speed. At a certain stage in the evolution of each device for transportation, limits have been reached beyond which he could not go. The ox-cart, the automobile, the airplane and the rocket, all have limits of speed which are basic and impossible to violate. The speed of the rocket, man's latest attempt, is limited by the velocity of the ejected gases, and this imposes upon the rocket a limitation of speed and range which man is reluctant to accept. In the coming age of space satellites and possible travel to the Moon, man will be casting envious eyes toward inter-planetary travel - travel into the depths of space where he may not even live long enough to complete his journey. It is already becoming apparent that the rocket must be superseded and speeds even further increased. The recognition of this obvious fact, even to rocket engineers, serves to dampen much of their enthusiasm about the practicability of travel by rocket spaceship. Fuel is consumed in "fighting" the gravitational field of the Earth. Fuel will be required in breaking the rate of fall, if and when landings are attempted on other planets. It is quite apparent that a method of controlling gravitation is urgently needed and "that it is already long overdue.

Two types of electrogravitic motors are proposed in Project WINTERHAVEN. Both types have a good chance of success. A motor weighing 500 lbs. for the propulsion of a model ship is suggested. Performance data derived from the tests of this model may be used in designing larger models, which in turn would presage electrogravitic motors for ocean liners weighing thousands of tons. Other possible applications, in due time, would include motive power for automobiles and rail-roads.

The second type of electrogravitic reactor now demonstrated in disc airfoils may find its principal field of usefulness in the propulsion of spaceships in various forms. For the moment, at least, the disc form appears to have the greatest promise, largely because there is reason to

111

believe it can be self-levitating and, therefore, made to possess the ability to move vertically (as well as horizontally) and to hover motionless, in complete control of the Earth's gravitational field.

Communications:

No person would have believed - if he had witnessed the original experiments of Prof. Hertz – that the obscure phenomenon would lay the groundwork for world-wide radio communication, radar, television and the countless electromagnetic devices of this kind which today we take so much for granted.

We have had, in our lifetime, the privilege of watching the growth and approaching culmination of the radio age. Yet, with all its manifest advantages, the electromagnetic wave has many limitations, and these are becoming increasingly apparent to us as over-crowded channels, annoying interference, blank-outs and shadows. We have become acutely aware of the troublesome limitations on television caused by the curvature of the earth and the shaded areas behind mountains, hills and large buildings, where satisfactory TV reception is virtually impossible. We sense that present methods are imperfect and inadequate and that somehow, in the future, an answer will be found.

If the basic experiments set forth in Project WINTERHAVEN prove the controllability of the gravitational wave, a fundamentally new system of communication will become available. Theory indicates that the gravitational wave may be one of the most penetrating forms of radiant energy. Employed as a means of communication, it may solve many of the difficulties inherent in present-day radio and, at the same time, provide countless additional channels for communication.

At the outset, development of the electrogravitational communication system obviously could provide a secret, almost wholly untouchable, channel for classified military communications. Message transmissions could be put through without breaking military radio silence, at a time when all electromagnetic transmissions are prohibited. Due to the high penetrability of the gravitational wave, communications could conceivably be maintained between submerged submarines, between submarines and shore installations or between bomb-proof shelters and similar underground installations without the use of external wires.

Other interesting possibilities virtually suggest themselves. Among these are the applications to undersea or under earth radar, also various remote control applications for guided missiles, where the usual antennae or dipole systems involve complications or create engineering difficulties because of the shielding of the metallic covering of the missiles.

Detection of distant atomic explosions:

Due to the tremendous momentary displacement of air and the gravitational disturbance resulting thereof, there is reason to believe that the electrogravitational receiver may be one of the few devices capable of instant long-distance detection and ranging of atomic bomb explosions.

Washington, D.C. Revised: 1/1/53

GENERAL OBJECTIVES

In the foregoing project outline, specific details have been referred to for the purpose of imparting a clear and concise understanding of the type of investigations proposed. The general objectives of Project WINTERHAVEN embrace the entire subject of the inter- relationships between gravitation and electrodynamics. This is necessarily a long-term program. Unquestionably there are many productive avenues of exploration in this vast and comparatively open territory which cannot be foreseen.

The project must adopt a policy of inviting suggestions from qualified physicists interested in attempting to solve the various problems involved. In a project of this scope and magnitude it would be a mistake to fail to recognize and investigate any phenomenon which bears even remotely upon the subject. It would be a mistake, for example, to limit the considerations to the so-called capacitor-effect, as outlined hereinbefore, when it's technical antithesis, a possible inductor-effect, may provide equal opportunities.

In the study of physical properties of dielectrics, low-temperature research is of especial importance. Electro-dynamic phenomena occur at low-temperatures which are completely unknown at room temperatures. The possibilities of discovering wholly unsuspected gravitational effects below the super-conductivity threshold, at temperatures approaching absolute zero, are worthy of the costs

113

involved. The use of the liquid helium cryostat is strongly recommended as an important part of Project WINTERHAVEN.

The operation of a library project such as that proposed for The Franklin Institute, for the accumulation of technical information and to serve as liaison with academic institutions throughout the world, is of utmost importance, particularly at the beginning of the program.

No responsibility can be assumed by any of the cooperating institutions to guarantee results in research. It is the express purpose of the sponsors of this project to seek the answers by organizing a cooperative program in which the best minds and all necessary laboratory facilities are brought together. It is the sincere hope that, in this way, a century of normal evolution in science, looking toward the ultimate control of gravitation for the benefit of mankind: may be compressed into 5-10 years.

As with the atomic bomb project in America, money was traded to gain time. So it is with the ultimate conquest of space. It must be recognized that a concentrated study of gravitation under a government research and development contract can no longer be neglected.

APPENDIX

PROGRAM OF FUNDAMENTAL RESEARCH

Section A. The Franklin Institute of the State of Pennsylvania
(a) Library project.-
(b) Liaison With other academic institutions.
(c) General considerations of field theories and gravitation
(d) Repetition of Trouton-Noble experiment with high-K dielectrics.
(e) Low temperature research of electrodynamic phenomena using liquid helium cryostat.

Section B. Stanford Research Institute.
(a) Repetition of Fernando Sanford experiments.
(b) Studies of variation in electrical potential of the Earth (c) Studies of electrogravitic induction.
(d) Development of a short-period gravimeter for capacitor tests.
(e) Cooperation with Lear, Inc., in studies of field relationships and gravimetric analysis.

Section C. Division of Statistical Analysis. National Bureau of Standards.
(a) Analysis of differential electrometer records.
(b) Analysis of capacitor mid point variations.
(c) Analysis of Sanford Variations.
(d) Correlations with solar, lunar and sidereal time.
(e) Correlations with other natural variables.

Section D. University of Chicago
(a) Repetition of basic pendulum experiment (Biefeld-Brown effect} in oil and other dielectric fluids, and in vacuum.
(b) Tests of ponderomotive forces in capacitors.
(c) Quantitative effects of K, m and other factors.
(d) Studies of high – K massive dielectrics and relation to forces developed.
(e) Consideration of inductor-effect in relation to condenser effect.
(f) Thrust measurements of electrified disc airfoils in air at reduced pressures, and in vacuum.

Antigravity on the Rocks: The T.T. Brown Story[41]

Jeane Manning

INTRODUCTION

T. Townsend Brown was jubilant when he returned from France in 1956. The soft-spoken scientist had a solid clue, which could lead to fuelless space travel. His saucer-shaped discs flew at speeds of up to several hundred miles per hour, with no moving parts. One thing he was certain, the phenomena should be investigated by the best scientific institutions. Surely now the science establishment would admit that he really had something. Although the tall, lean physicist—handsome, in a gangly way—was a humble man, even shy, he confidently took his good news to a top-ranking officer he knew in Washington, D.C.

"'The experiments in Paris proved that the anomalous motion of my disc airfoils was not all caused by ion wind." The listener would hear Brown's every word, because he took his time in getting words out. They conclusively proved that the apparatus works even in high vacuum. Here's the documentation.

Anomalous means unusual—a discovery which does not fit into the current box of acknowledged science. In this case, the anomaly revealed a connection between electricity and gravity.

That year, *Interavia* magazine responded that Brown's discs reached speeds of several hundred miles per hour when charged with several hundred thousand volts of electricity.[42] A wire running along the leading edge of each disc charged that side with high positive voltage, and the trailing edge was wired for an opposite charge. The high voltage ionized air around them, and a cloud of positive ions formed ahead of the craft and a cloud of negative ions behind. The apparatus was pulled along by its self-generated gravity field, like a surfer riding a wave. *Fate* magazine writer, Gaston Burridge in 1958 also described Brown's metal discs, some up to 30 inches in diameter by that time.[43] Because they needed a wire to supply electric charges, the discs were tethered by a wire to a Maypole-like mast. The double-saucer objects

---

[41] Another version of this article also appears in *Suppressed Inventions & Other Discoveries* by J. Eisen, Penguin Pub. – Ed note

[42] See *Interavia* article, perhaps the last media exposure on gravity, on page 62 – Ed. note

[43] See "Another step toward anti-gravity" by Gaston Burridge, *The American Mercury*, 86(6) 1958, p.77-82

circled the pole with a slight humming sound in the dark they glow with an eerie lavender light."

Instead of congratulations on the French test results, at the Pentagon he again ran into closed doors. Even his former classmate from officers' candidates School, Admiral Hyman Rickover, discouraged Brown from continuing to explore the dogma-shattering discovery that the force of gravity could be tweaked or even blanked out by the electrical force.

"Townsend, I'm going to do you a favor and tell you: Don't take this work any further. Drop it."

Was this advice given to Brown by a highly-placed friend who knew that the United States military was already exploring electrogravitics? (Recent sleuthing by American scientist, Dr. Paul La Violette, uncovers a paper trail which leads from Brown's early work, toward secret research by the military and eventually points to "Black Project" aircraft.)

HARASSMENT

Were the repeated break-ins into Brown's laboratory meant to discourage him from pursuing his line of research?

Brown didn't quit, although by that lime he and his family had spent nearly $250,000 of their own money on research. He had already put in more than thirty years seeking scientific explanations for the strange phenomena he witnessed in the laboratory. He earlier called it electrogravitics, but later in his life, trying to get acknowledgment from establishment scientists, he stopped using the word "electrogravitics" and instead used the more acceptable scientific terminology "stress in dielectrics".

No matter what his day job, the obsessed researcher experimented in his home laboratory in his spare time. Above all he wanted to know, "Why is this happening?" He was convinced that the coupling of the two forces – electricity and gravity could be put to practical use.

An arrogant academia ignored his findings. Given the cold-shoulder treatment by the science establishment, Brown spent family savings and even personal food money on laboratory supplies. Perhaps he would not have had the heart to continue his lonely research if he had known in 1956 that nearly 30 more years of hard work were ahead of him. He died in 1985 with the frustration of having his findings still unaccepted.

The last half of his career involved new twists. Instead of electrogravitics, at the end of his life he was demonstrating "gravitoelectrics" and "petrovoltaics"— electricity from rocks. Brown's

117

many patents and findings ranged from an electrostatic motor to unusual high-fidelity speakers and electrostatic cooling, to lighter-than-air materials and advanced dielectrics. His name should be recognized by students of science, but instead it has dropped into obscurity.

Too late to comfort him, some leading-edge scientists of the mid-1990s are now resurrecting Brown's papers. Or what they can find of his papers.

EXTRAORDINARY CURIOSITY

Thomas Townsend Brown was born March 18, 1905, to a prominent Zanesville, Ohio, family. The usual child-like "Why?" questions came from young Townsend with extraordinary intensity. For example, his question "Why do the (high voltage) electric wires sing?" led him later in life to an invention.

His discovery of electrogravitics, on the other hand. came through an intuition. As a sixteen-year-old, Townsend Brown had a hunch that the then-famous Coolidge X-ray tube might give a clue to spaceflight technology. His tests, to find a force in the rays themselves, which would move mass, lead to a dead end. But in the meantime the observant experimenter noticed that high voltages applied to the tube itself caused a very slight motion.

Excited, he worked on increasing the effect. Before he graduated from high school, he had an instrument he called a gravitator " Wow," the teenager may have thought., "Antigravity may be possible!" World-changing technological discoveries start with someone noticing a small effect and then amplifying it.

Unsure of what to do next, the next year he started college at California Institute of Technology. Even then his sensitivity was evident because he saw the wisdom of going forward cautiously-first gaining respect from his professors instead of prematurely bragging about his discovery of a new electrical principle. He was respected as a promising student and an excellent laboratory worker, but when he did tell his teachers about his discovery they were not interested. He left school and joined the Navy.

Next he tried Kenyon College in Ohio. Again, no scientist would take his discovery seriously. It went against what the professors had been taught, therefore it could not be.

He finally found help at Dennison University in Gambier, Ohio. Townsend met Professor of physics and astronomy Paul Alfred Biefeld, Ph.D., who was from Zurich, Switzerland and had been a classmate of Albert Einstein. Biefeld encouraged Brown to experiment

118

further, and together they developed the principle which is known in the unorthodox scientific literature as the Biefeld-Brown Effect. It concerned the same notion which the teenager had seen on his Coolidge tube–a highly charged electrical condenser moves toward its positive pole and away from its negative pole. Brown's gravitator measured weight losses of up to one percent. (In 1974 researcher Oliver Nichelson pointed out to Brown that before 1918, Professor Francis E. Nipher of St. Louis discovered gravitational propulsion by electrically charging lead balls, so the Brown-Biefeld Effect could more properly be called the Nipher Effect. However, Brown deserves credit for his sixty years of experimentation and developing further aspects of the principle.)

Brown's 1929 article for the publication *Science and Invention* was titled bluntly, "How I Control Gravity."[44] The science establishment still turned its back. By then he had graduated from the university, married, and was working under Professor Biefeld at Swazey Observatory. His career in the early 1930's also included a post at the Naval Research Lab in Washington, D.C. as staff physicist for the Navy's International Gravity Expedition to the West Indies; physicist for the Johnson-Smithsonian Deep Sea Expedition; and soil engineer for a federal agency and administrator with the Federal Communications Commission.

As his country's war effort escalated, he became a Lieutenant in the Navy Reserve and moved to Maryland as a materials engineer for the Martin Aircraft Company. Brown was then called into the Navy Bureau of Ships. He worked on how to degauss (erase magnetism from) ships to protect them from magnetic-fuse mines, and his magnetic minefield detector saved many sailors' lives.

PHILADELPHIA EXPERIMENT

The "Philadelphia Experiment" which Brown may or may not have joined in 1940 is dramatized in a popular movie as a military experiment in which United States Navy scientists are trying to demagnetize a ship so that it will be invisible to radar. According to the account, the ship and its crew dematerialized and rematerialized – became invisible and later returned from another dimension.

Whatever the Project Invisibility experiment actually was, Brown was probably an insider, as the Navy's officer in charge of magnetic and acoustic mine-sweeping research and development. However, later in life, Brown was said to be mute on the topic of the alleged

---

[44] "How I Control Gravity" article appears on page 56 – Ed. note

Philadelphia Experiment, except for brief disclaimers. He told Josh Reynolds of California, who made arrangements for Brown's experiments in the early 1980's, that the movie and the controversial book *The Philadelphia Experiment*, by William L. Moore and Charles Berlitz, were greatly inflated.[45] He apparently didn't elaborate on that comment.

Reynolds spoke on a panel discussion at a public conference (dedicated to Townsend Brown) in Philadelphia in 1994, along with highly-credentialed physicist Elizabeth Rausher, Ph.D. Rauscher theorized that the Philadelphia Experiment legend grew out of the fact that certain magnetic fields can in effect "degauss the brain" – cause temporary memory loss. If the huge electrical coils involved in degaussing a ship were mistuned, the sailors could have felt that they "blinked out of time and back into time."

Blinking this account back to 1942: Townsend Brown was made commanding officer of the Navy's radar school at Norfolk, Virginia. The next year he collapsed from nervous exhaustion and retired from the Navy on doctors' recommendations. More than his hard work caused his health to break down. He had suffered years of deeply-felt disappointments because his life's work—the gravitator—had not been recognized by scientific institutions which could have investigated it. The final precipitating factor for his collapse was an incident involving one of his men.

BREAK-IN AT PEARL HARBOR

After he recuperated for six months, his next job was as a radar consultant with Lockheed-Vega. He later left the California aircraft corporation, moved to Hawaii and was a consultant at the Navy yard at Pearl Harbor. An old friend who was teaching calculus there had opened some doors, and in 1945 Brown demonstrated his latest flying tethered discs to a top Military officer—Admiral Arthur W. Radford, commander-in-chief for the U.S. Pacific Fleet, who later became Joint Chief of Staff for President Dwight Eisenhower.

Brown was treated with respect because of who he was, but again no one signed up to help investigate his discovery. His colleagues in the Navy treated it lightly because it was anomalous.

When he returned to his room after the Pearl Harbor demonstration, however, the room had been broken into and his notebooks were gone. A day or so later, as Josh Reynolds remembers Brown's account of the incident, "they came to him and said 'we have your work; you'll get it

---

[45] Ballantine Books, 1979

back.' A couple of days later they gave him back his books and said, 'we're not interested.'"

"Why?" Brown was given the answer that the effect was a result of ion propulsion, or electric wind, and therefore could not be used in a vacuum such as outer space, The earth's atmosphere can be rich in ions (electrically-charged particles), but a vacuum is not.[46]

He was disgruntled, but not stopped. Later a study funded by a French Government agency would prove the effect was not caused by "electric wind." But even before that Brown knew that it would take an electric hurricane to create the lifting force he saw in his experiments.

Project Winterhaven was his own effort for furthering electrogravitic research.[47] He began the project in 1952 in Cleveland, Ohio. Although he demonstrated two-feet-diameter disk-shaped transducers which reached a speed of 17 feet per second when electrically energized, he was again met with lack of Interest. Alone in his enthusiasm, he watched the craft fly in a 20-foot diameter circle around a pole. According to the known laws of science, this should not be happening. And he went on to make spectacular demonstrations.

When La Societe Nationale de Construction Aeronautique Sud Quest (SNCASO) in France offered him funding, he went to France and built better devices as well as had them properly tested. Those tests convinced his backers that it could mean a feasible drive system for outer space, he told Reynolds. SNCASO merged with Sud East in 1956 and funding was cut, so Brown had to return to the US.

Brown was eager to show the French documentation to all those officials who had raised the wall of indifference in the past. But after his discouraging visit to Washington, D.C. in 1956 and what felt like a put-down from Admiral Rickover, he apparently decided "if the military isn't interested, the aerospace companies might be." Friends say it did not occur to him to ask if the defense industry was already working on electrogravitics, unknown to him. In 1953, he had flown saucer-shaped devices of three feet in diameter in a demonstration for some Air Force officials and men from major aerospace companies. Energized with high voltage, they whizzed around the 50-foot diameter course so fast that the reports of the test were stamped "classified." (In the transcribed telephone conversation recorded on the first page of the ONR report, Gen. Bertrandias tells Gen. Craig three

---

[46] See the comprehensive evaluation by the Office of Naval Research entitled, "*The Townsend Brown Electro-Gravity Device*" 15 September 1952, File 24-185, 28-page report, reprinted in its entirety by Integrity Research Institute

[47] See T.T. Brown's Project Winterhaven proposal on p. 90

times how "frightened" he was about having flying saucer tests "conducted by a private group.")[48]

Independent researcher, Paul LaViolette, Ph.D. traces the path which these impressive results led toward the Pentagon, the military hub of the United States. A recently declassified Air Force intelligence report indicates that by September of 1954 the Pentagon had launched a program to develop a manned antigravity craft of the sort suggested in Project Winterhaven, writes La Violette.[49]

Meanwhile, Brown went practically door-to-door in Los Angeles to try to rouse some interest in his work. One day he returned to his laboratory to find it had been broken into and much of his belongings were missing.

CHARACTER ASSASSINATION

Then the nasty rumors started. The type of rumors, which can discredit a man's character, upset his wife and children and overall cause deep distress to a sensitive man.

Another tragedy in Brown's life was the sudden death of his friend and helpful supporter, Agnew Bahnson. who funded him to do anti-gravity research and development beginning in 1957 in North Carolina. Did they make too much progress? In 1964 Bahnson, an experienced pilot, mysteriously flew into electric wires and crashed. Bahnson's heirs dissolved the project.

The authors of the book *The Philadelphia Experiment* wrote that "in spite of his numerous patents and demonstrations given to governmental and corporate groups success eluded Townsend Brown. Such interest as he was able to generate seemed to melt away almost as fast as it developed almost as if someone was working against him."

Today's researchers looking at Townsend Brown's life have noticed that he went into semi-retirement some time in the 1960s. Tom Valone, of Washington, D.C., who in 1994 compiled a book on Brown's work,[50] speculates that the work was classified and Brown was bought off or somehow persuaded to stop promoting electrogravitics. Valone told the April, 1994 meeting in Philadelphia that Dr. LaViolette's detective work sheds new light on what happened to Brown in the 1950s. The speculation of these scientists is that "this

---

[48] See Brown's flying saucer demonstration unit on p. 133 – Ed note

[49] See LaViolette's article, "The US Antigravity Squadron" – reprinted in *Electrogravitics Systems,* the first volume in this series – Ed. note

[50] *Electrogravitics Systems: A New Propulsion Methodology*, 1st edition, Integrity Research Institute, 1994 – Ed. note

project was taken over by the military, worked on for 40 years, and we now have a craft that's flying around." Valone speculates that Brown was debriefed and told what he could talk about.

In the later 1960s to 1985, Brown turned his attention to other research, although related. He mainly did basic research to try to understand strange effects he saw. As did T. H. Moray, Townsend Brown had decided that waves corning from outer space are not only detected on Earth, but also the waves build up a charge in a properly built device. Instead of making increasingly complex devices, however, Brown toward the end of his life in the 1980s was getting a charge—voltage to be exact—out of rocks and sand.[51] It was all in search for answers.

If his work had been accepted instead of suppressed by seeming disinterest, he would be known to science students. His work would fill more than one science book: an encyclopedia set could easily be filled with T. T. Brown's experiments and discoveries.

For example, his childhood fascination with the singing wires led him to investigate how to modulate ionized air like that which had carried the high-voltage current. Could this be, used for high-fidelity sound systems? Eventually he did invent rich-sounding Ion Plasma Speakers which incidentally had it built-in "fac" – a cool breeze of health-enhancing negative ions. Would this discovery have been commercialized if his main interest, electrogravitics, had not been suppressed by ignorance or been co-opted?

He searched for better dielectrics endlessly trying new combinations. (A dielectric is any material, which opposes the flow of electric current while at the same time can store electrical energy.) This search led him to study, when working with Bahnson, the lighter-than-air fine sand, in certain dry river beds, which could be used to make advanced materials. The anomalous sands were first discovered by his hero Charles Brush early in the century. Brush also found that certain materials fell slower in a vacuum chamber than others. He called it gravitational retardation and said they were slightly more interactive with gravity. These materials also spontaneously demonstrated heat. Brush believed that the "etheric gravitational wave" interacted with some materials more richly than with others. Brush's findings were swept under the rug of the science establishment,

Brown followed his idol's lead and did basic research in a number of area." Gravito-electrics—how neutrinos or gravitons or whatever-they-are converted into electricity. This led him to conduct experiments in

---

[51] It is called "petrovoltaics" – Ed. note

various locations, from the ocean to the bottom of the Berkeley mineshaft. When entrepreneur Josh Reynolds became interested in Brown's work in the last five years of the inventor's life, Brown was able to do the work he loved the most—petrovoltaics. No one else was putting electrodes on rocks to measure the minute voltages of electricity which the rocks some-how soaked up from the cosmos. Brown and Reynolds made artificial rocks to see what various materials could do and how long they would put out a charge.

Their efforts in a number of areas led toward what they called a Forever Ready Battery—a penny-sized piece of rock which put out a tiny amount of voltage indefinitely because they had learned how to "soup-up" the effect. After Brown died, Reynolds carried on the research until funding ran out. He estimates that it would have taken up to $10 million of advanced molecular engineering research to take the discovery to another stage of development. The high-power version of the battery remains on paper-only theory until developed farther.

This discovery alone should have put Brown into science history books. In all his years of experiments with the periodic variations in the strip-chart recordings of the output from the materials, he found that the patterns had a relationship to position of the stars. And orientation toward the center of the universe seemed to make a difference too. This resulted in further unconventional thinking that only made Brown more of an out-cast in the world of sanctioned science.

While he was coming up with the cosmic findings, the military researchers had a different agenda. One of the reports dug up by researcher LaViolette came from a London think tank called Aviation Studies International Ltd. In 1956 the think tank wrote a "confidential" survey of work done in electrogravitics. LaViolette says the only original copy of the document, called Report 13, was found in the stacks at Wright-Patterson Air Force Base technical library in Dayton, Ohio. It is not listed in the library's computer.[52]

Excerpts from Report 13 paint a picture of heavy secrecy. A 1954 segment says that infant science of electro-gravitation may be a field where not only the methods are secret but also the ideas themselves are a secret. "Nothing therefore can be discussed freely at the moment." A further report predicted bluntly that electrogravitics, like other advanced sciences, would be developed as a weapon.

A couple of months later, another now declassified Aviation Report said it looked like the Pentagon was ready to sponsor electrogravitic

---

[52] The first report in the book: "Electrogravitics Systems" – Ed. note

propulsion devices and that the first disc should be finished by 1960. The report anticipated that it would take the decade of the 60s to develop it properly "even though some combat things might be available ten years from now."[53]

Defense contractors began to line up, as well as universities who get grants from the U .S. Department of Defense. After he came across Report 13, LaViolette put his knowledge of physics to work and began to piece together a picture of what may have happened in the past thirty years. It includes "black" projects—work which the military decides should be so secretive that even Congress does not get reports about its funding.

A breakthrough in LaViolette's quest for the pieces of the picture came when a few establishment scientists gave out tidbits of formerly-secret Information about a "black funding" project—the Stealth B-2 bomber. (The B-2 is described as the world's most expensive aircraft at $1.2 billion.) Their description of the B-2 gave LaViolette and others a number of clues about the bomber-softening of the sonic boom as Brown had talked about in the 1950s, a dielectric flying wing, a charged leading-edge, ions dumped into the exhaust stream and other clues. The B-2 seems to be a culmination of many of Brown's observations made more than forty years ago.

Townsend Brown fought an uphill battle all his adult life, at great cost to himself and to family life. His cause included getting the science of advanced propulsion out into public domain, not hidden behind the Secrecy Act and a wall of classified documents. He died feeling that he had lost the battle.

Thomas Townsend Brown

---

[53] The second report in this book: "The Gravitics Situation" – Ed. note

TESTIMONIAL SECTION

Edge-on photo of the **B-2 Bomber** which no longer displays the black dielectric layer on the leading edge of the wings, as seen in *Electrogravitics Systems*, Figure 1, p.79. Photo courtesy of Northrop Grumman.

Email from Richard Boylan, Ph.D.

I'm forwarding Bobbi Garcia's interesting photo of the B- 2 Stealth Bomber in flight. Since it is now known that the B-2 has electrogravitics on board (with gravity-cancelling properties of approximately 89% or higher efficiency, utilizing a leak from defense contract insider Edgar Rothschild-Fouche, who descibed a similar system on the TR3-B antigravity triangle-plane,) it is to be noted the high coronal discharge around the airframe once it switches from take-off conventional jet turbine propulsion to electro-gravitic field propulsion.

Bobbi shot a fabulous photo of the B-2 high speed, low level, when she was flying chase on a recent mission.[54] Her shot was awarded lst place in *Aviation Week & Space Technology's* Military Category and appeared in the 23 Dec 2002 issue.

Photo by Boeing employee Bobbi Garcia.

---

[54] This shock wave also follows the electrogravitic field shape as diagrammed by Dr. Paul LaViolette on p. 90 of Volume I   - Ed note

Correspondence Relevant to the B-2 Bomber Paper[55]

Date: Mon, 22 Sep 1997 17:14:07 -0500 (CDT)
From: rich.boylan@24stex.com
Subject: More on B-2 Stealth bomber as antigravity craft

Retired Air Force Colonel Donald Ware has passed on to me information from a three-star general he knows who revealed to him in July that "the new Lockheed-Martin space shuttle [National Space Plane] and the B-2 [Stealth bomber] both have electro-gravitic systems on board;" and that " this explains why our 21 Northrop B-2s cost about a billion dollars each." Thus, after taking off conventionally, the B-2 can switch to antigravity mode, and, I have heard, fly around the world without refueling.

March 26,1999, Brooklyn, NY
Alex Cavallari, xelaufo@aol.com reported:

While watching an early morning TV news show, "Good Day NY" on FOX channel 5 in NY at about 8:12 - 8:14 AM est on 3-26-1999 there was a segment on called "Crisis in Kosovo." The guest for this segment was a Mr. Cliff Bragdon, he claims to be a military and defense expert, while being questioned on the aircraft and technology being used by the USA military in Kosovo, Mr. Cliff Bragdon stated that the Stealth B2 Bomber "uses antigravity technology".

---

[55] Reference is made to "The US Antigravity Squadron" by Dr. Paul LaViolette, which analyzes the B-2 bomber in detail, including the auxiliary power system, reprinted on p. 78 of the Volume I book, *Electrogravitics Systems.* – Ed note

Testimony of Dr. B., December, 2000

Excerpt from *Disclosure* by **Steven Greer, MD**,
Page 269-270

...A few years ago a guy came down from Moffett Field, NASA Ames and he took me to dinner. He showed me his Government card and everything. He said, we have a little $50,000 grant we want to give you. I said, well, that's pretty interesting. I haven't done anything for NASA in a long time. He says, all we want you to do is come up with an idea to cut down wind resistance on commercial airliners to get better fuel economy out of a jet. I said, fine. I'll do that.

This is also in my book. I have pictures of it. So I showed him a design. I took a 737, and said let's make the engine a flame jet generator, because it's a very great source of static energy. There are millions of ergs of electricity being wasted from that thing. We'll hook that up. We'll put a positive charge across the front of the plane, the wings. We'll put a negative charge on the trailing surfaces. We'll do the rudder, the elevator, and the forward part of the wings. We'll use Mylar as insulation. We'll use platinum rhodium plates and we'll put a big positive charge. The faster the thing goes, the more energy it puts out. It'll shoot out a positive charge of particles out in front in a trail, so it'll cut the wind resistance to almost nothing in this plane.

It'll start working around 200, 250 knots, just after the thing does take off: it'll be at V-3. When it gets to altitude, it'll be awesome. I sent him some drawings. And the guy calls me back in a week and says, Mr. B, this is way beyond what we wanted from you. He says we can't do this. We can't do this. I said, why not? It'll work. He says, yes it'll work, but we didn't want anything this technical. Well, I realized this is really weird, this conversation. (I should show you his card before you leave.)

So then, I talked to my friends over at *Aviation Weekly*, Mark McCandlish, and I found out what I had just done was to design the front-edge of a B-2 Bomber, which goes supersonic.[56] I had just given them the design to what they already had, and it freaked them out, because I gave them a classified design, which I had from the Lockheed Skunk Works.

That came from my buddy over at Lockheed Skunk Works who finally disappeared, by the way. He got started talking a lot, and he disappeared. He's not around any more. He just disappeared one day. Nobody knows where he went. His place was closed. Overnight he was gone. Yeah, he was a great contact. He told me all about the Aurora.

---

[56] See *Electrogravitics Systems: Reports on a New Propulsion Methodology* for more detailed information about the B-2 electrogravitics design – Ed. note

Testimony of Mr. Mark McCandlish, December, 2000
Excerpt From *Disclosure* by **Steven Greer, MD**

Mark McCandlish is an accomplished aerospace illustrator and has worked for many of the top aerospace corporations in the United States. His colleague, Brad Sorenson, with whom he studied, has been inside a facility at Norton Air Force Base, where he witnessed alien reproduction vehicles or ARVs that were fully operational and hovering. In his testimony, you will learn that the US not only has operational antigravity propulsion devices, but we have had them for many, many years and they have been developed through the study, in part, of extraterrestrial vehicles over the past fifty years. In addition, we have the drawing from aerospace inventor Brad Sorenson of the devices that he saw, as well as a schematic of one of these alien reproductions vehicles—in some remarkable detail.

I work principally as a conceptual artist. Most of my clients are in the defense industry. I occasionally work directly for the military, but most the time I work for civilian corporations that are defense contractors and build weapons systems and things for the military. I've worked for all the major defense contractors: General Dynamics, Lockheed, Northrop, McDonald-Douglas, Boeing, Rockwell International, Honeywell, and Allied Signer Corporation.

In 1967 when I was at Westover Air Force Base, one night before going to bed I saw this light moving across the sky; then it just kind of stopped, and there wasn't any noise. I took the dog back in the house, and I brought out my telescope and watched this thing through the telescope for about ten minutes. In fact, it was hovering directly over the facility where they kept the nuclear weapons - at the storage facility near the alert hangers at Westover Air Force Base. It started to move off, and it moved off slowly and kind of wandered around the sky. Then, all of a sudden it was gone, like it had been fired out of a gun. It was out of sight in just a second or two.

Well, it all started coming together when I was working at IntroVision, and John Eppolito talked about this interview that he had done with a person who had, for some reason, wound up walking up to, or near a hangar at a section of a military Air Force base. [He] had seen a flying saucer in a hangar, and then he was arrested—hauled off, blindfolded, and debriefed – this sort of thing. Then I learned that this fellow, Mark Stambough, had developed an experiment that created a kind of levitation. In some circles it's been called electrogravitic levitation, or antigravity.

What he did, apparently, was acquire a high voltage power source—a DC (direct current) power source, and he took a couple of quarter-inch-thick copper plates about a toot in diameter, with a lead coming out of the middle of each one at the top and the bottom. [Then], he basically embedded them in a type of plastic resin like polycarbonate or Plexiglas, or some other kind of clear resin where you could see the plates, and you could see the material. Apparently, he did everything he could to get all the little bit bubbles and stuff out of there, so there wouldn't be any pathways for the electricity to break down the material and arc through them. The experiment – as to see how much voltage you could put on this capacitor – the sub-plate capacitor – in this arrangement; how much voltage could you put on this thing before the insulating material begins to just break down?

Well, he got up to about a million volts, and the thing would begin to float, and it floated in accordance with principles that had been described in a patent that was filed back in the late 1950s/early 1960s by a gentleman called Thomas Townsend Brown. Brown and another individual by the name of Dr. Biefield had done this, so this effect became known as the Biefield-Brown effect. Well, [Stambough] apparently duplicated the experiments done by Biefield and Brown, [and] the one aspect they found about this arrangement was that the levitation or movement would occur in the direction of the positively-charged plate. So, if you had two plates, one is negative, and one is positive because of the direct current system. If you have the positive plate on top, it would move in that direction. If you had it on a pendulum, it would always swing in whatever direction the positive plate was facing.

Later, I got a call from a kid that I had known in school, a fellow by the name of Brad Sorensen. He apparently had recognized my name [from some work I had done in a magazine], and had contacted the art director who gave him my phone number, and he called me up. It turned out that he had gone to work for a design firm in the Glendale/Pasadena area of California and ultimately wound up acquiring most of the clientele for this particular agent.

In the process, he developed a business practice where he would create conceptual designs and products for different clients. The way he structured his business [was to] set it up so that if he came up with some new and novel design, something that was patentable, the client would pay to have the patent secured. Then he would agree, if the patent was in his name, to license it exclusively to them and no one else, and they would pay him royalties. So, he got his clients to pay for all these patents, and he had all these royalties coming in, and he was a millionaire before 30.

So, this is Brad Sorensen coming back to me eight years after school, and we're talking, and he's telling me all these interesting stories. There was an air show that was coming up at Norton Air Force Base, which used to be an active Air Force base right on the eastern fringe of San Bernardino in Southern California.

I suggested that we get together and go to this air show. I had heard that they were going to have a fly-by (a flying demonstration) by the SR-71 Blackbird, and he seemed to know a lot about it, so I said, well, let's do that. It turned out [that] at the last minute, the magazine *Popular Science* came back again and [told me] they had some really, really crazy deadline for another illustration, and they wanted to know if I could do it over the weekend, so I had to beg off on this air show. Brad had already made arrangements to go, and he was going to bring one of his clients with him. It turned out that this client was a tall, thin, white-haired man with glasses [and] an Italian-sounding last name. He was already a millionaire in his own right and was in civilian life again after having been either a Secretary of Defense or an Under-Secretary of Defense. Brad wanted me to meet this gentleman, and if I had known this at the time, I probably would have told the magazine to wait, because I had no idea at that point what I was going to be missing out on.

Believe me, I've kicked myself ever since, because the following week, after Brad got back home, he called me up and told me about the air show. He told me about what he had seen there: apparently, right about the time the Air Force flying demonstration team, the Thunderbirds, were planning to begin their show, this gentleman that Brad was with said, "Follow me," and they [went] walking down to the other end of the airfield, away from where the crowds were, to this huge hangar that's at Norton Air Force Base. I don't remember the building number, but it's got to be one of the largest hangars in the Air Force inventory.

In fact, on the base it was called 'The Big Hangar.' It looks like four giant Quonset hut style hangars that are all connected in the middle, with shops and work areas out around the edges, and there is sort of a divider in the middle.

This gentleman took Brad down there. He said, "I'm here to talk to the guy who is running the show," so the guard goes in and out comes the same guard with a gentleman in a three-piece suit, who immediately recognizes this fellow that Brad is with: this fellow whom I speculate was probably Frank Carlucci. They go inside, and immediately after getting inside the door, this fellow apparently passes Brad off as his aide to this fellow who is managing the exhibit that's going on inside this hangar. This exhibit is for some of the local politicians who are cleared for high security information, [plus] some of the local military officials.

Well, as soon as they walk in, Brad is told by this fellow that he is with, "There are a lot of things in here that I didn't expect they were going to have on display- stuff you probably shouldn't be seeing. So, don't talk to anybody, don't ask any questions, just keep your mouth shut, smile and nod, but don't say anything – just enjoy the show. We're going to get out of here as soon as we can."

In the process, the host or the person running the show was very engaging with the gentleman that Brad was with, so they bring them in, and they are showing them everything. There was the losing prototype from the B-2 Stealth Bomber competition. They also had what was called the Lockheed Pulsar, nicknamed the Aurora.

These things had the ability to be just about anywhere in the world 30 minutes after launch, with the capability of 121 nuclear warheads – you know, probably 10-15 megaton weapons – a tactical type nuclear reentry vehicle.

So, getting back to Brad's story at Norton Air Force Base: one of the other things he said was that after they showed them all of these aircraft, they had a big black curtain that divided the hangar into two different areas. Behind these curtains was another big area, and inside this area they had all the lights turned off; so, they go in and they turn the lights on, and here are three flying saucers floating off the floor-no cables suspended from the ceiling holding them up, no landing gear underneath-just floating, hovering above the floor. They had little exhibits with a videotape running, showing the smallest of the three vehicles sitting out in the desert, presumably over a dry lakebed – someplace like Area 51. It showed this vehicle making three little quick, hopping motions; then [it] accelerated straight up and out of sight, completely disappearing from view in just a couple of seconds-no sound, no sonic boom-nothing.

They had a cut-away illustration, pretty much like the one I'll show you in a little bit, that showed what the internal components of this vehicle were, and they had some of the panels taken off so you could actually look in and see oxygen tanks and a little robotic arm that could extend out from the side of the vehicle for collecting samples and things. So, obviously, this is a vehicle that not only is capable of flying around through the atmosphere, but it's also capable of going out to space and collecting samples, and it's using a type of propulsion system that doesn't make any noise. As far as he could see, it had no moving parts and didn't have any exhaust gases or fuel to be expended – it was just there hovering.

So, he listened intently and collected as much information as he could, and when he came back, he told me how he had seen these three flying saucers in this hangar at Norton Air Force Base on November 12, 1988 – it was a Saturday. He said that the smallest was somewhat bell-shaped. They were all identical in shape and proportion, except that there were three different sizes. The smallest, at its widest part, was flat on the bottom, somewhat bell-shaped, and had a dome or a half of a sphere on top. The sides were sloping at about a 35-degree angle from pure vertical.[57]

The panels that were around the skirt had been removed, so he could see one of these big oxygen tanks inside. He was very specific in describing the oxygen tanks as being about 16 to 18 inches in diameter, about 6 feet long, and they were all radially-oriented, like the spokes of a wheel. This dome that was visible on the top was actually the upper half of a big sphere-shaped crew compartment that was in the middle of the vehicle, and around the middle of this vehicle was actually a large plastic casting that had this big set of copper coils in it. He said it was about 18 inches wide at the top, and about 8 to 9 inches thick. It had maybe 15 to 20 stacked layers of copper coils inside of it. The bottom of the vehicle was about 11 or 12 inches thick. In both cases, the coil and this large disc at the bottom were like a big plastic casting – sort of a greenish-blue, clear plastic, or it might have been glass. I determined, using my conceptual artist skills, that there were exactly 48 sections like thin slices of pizza pie, and each section within this casting probably weighed four or five tons, judging by the thickness and the diameter. It must have been monstrous in weight. It was full of half-inch-thick copper plates, and each of the 48 sections had 8 copper plates.

So, here we are back to the plate capacitors again, and the prospect of someone finding a way to use the Biefield-Brown effect – this levitation effect where you charge a capacitor to lift towards a positive plate. Now, when you've got eight plates stacked up in there, they alternate. It goes: negative positive, negative, positive, negative, positive – four times, so you ultimately wind up with the positive plates always being above a set of negative plates as you go up.

On the inside of the crew compartment was a big column that ran down through the middle, and there were four ejection seats mounted

---

[57] See Mark's detailed drawing at the end of this article   – Ed note

back-to-back on the upper half of this column. Then, right in the middle of the column, was a large flywheel of some kind. Well, this craft was what they called the Alien Reproduction Vehicle; it was also nicknamed 'the Flux Liner.' This antigravity propulsion system – this flying saucer – was one of three that were in this hangar at Norton Air Force Base. [Its] synthetic vision system [used] the same kind of technology as the gun slaving system they have in the Apache helicopter: if [the pilot] wants to look behind him, he can pick a view in that direction, and the cameras slew in pairs. [The pilot] has a little screen in front of his helmet, and it gives him an alternating view. He [also] has a little set of glasses that he wears – in fact, you can actually buy a 3-D viewing system for your video camera now that does this same thing-so when he looks around, he has a perfect 3-D view of the outside, but no windows. So, why do they have no windows? Well, it's probably because the voltages that were talking about [being] used in this system were probably something between, say, half a million and a million volts of electricity.

Now, he said there were three vehicles. The first one – the smallest, the one that was partially taken apart, the one that was shown in the video that was running in this hangar November 12, 1988 at Norton Air Force Base – was about 24 feet in diameter at its widest part, right at the base. The next biggest one was 60 feet in diameter at the base.

Now, I started looking at the design of this thing, and it occurred to me that what I was looking at was a huge Tesla coil, which is kind of like an open-air transformer. What happens is that when you pass electricity through this large diameter coil, it creates a field.

That's what this system does: it takes electricity, using two large 24-volt marine-style batteries. You basically use that to somehow put an alternating current through these windings. [Then], you step up that electricity through the secondary coil, which is on the column in the middle, and you get this extremely high voltage. You can selectively put the voltage on any of these 48 capacitor sections.

Well, why would you want to do that? If you're using just a normal Tesla coil, you usually have maybe one or two capacitors in the whole system. But, you're talking about a different type of capacitor here – you're talking about capacitors that are made up of plates-plates that are shaped like long, thin triangles, and they are all radially-oriented just like the spokes of a wheel, just like the oxygen tank, just like the field lines from that large diameter coil. As you look at this system, if you're an electrician or just somebody who knows a little bit about Tesla coils and the way they are set up, you begin to realize that the orientation of components is really the key to making the system work.

Why so many different capacitor sections? If you just have one big disc like. Mark Stambough did with his experiment at the University of [Arizona] – which, by the way, was confiscated by men claiming to be from the government and claiming privilege under the National Security Act. They took all this stuff, interviewed all the people that saw the experiments, and told everybody to keep their mouths shut and not talk about it. But, I heard about it from his roommate who knew what had happened. [Anyway], in that case, you have levitation, but you don't get any control. You have this thing floating around, and it's just sort of floating on whatever this field is that it's producing, but you don't have any control.

So, what happens? You break that disc up into 48 different sections, and then you can decide how much electricity you want to put on this side or over there on that side, so you can control the amount of electricity and the amount of thru and vectoring that you get. You can make it go straight up; you can make it bank and turn and pitch-whatever you want to, by virtue of the fact that you can control where the electricity goes in those 48 different sections. If you ever take a circle and divide it up into 48 equal sections, you'll find that those are really thin little slices. So, you have these 48 individual capacitors, and you have one big Tesla coil. You've got to have some kind of a rotating spark gap, just like the distributor in your car that sends the electricity out to each of those sections. Then, you have to have some way of controlling how much electricity goes to each one.

[A disc-shaped craft like this has omni-directional movement – it isn't limited to moving in one direction like a jet with a nose and a tail. *LW after talking to McCandlish.*]

Now, when Brad described the control system, he said that on the one side there was this big high-voltage potentiometer – it's like a rheostat, a big controller. It allows you to put progressively more electricity through the system as you push the lever. On the other side of the control system, there was a sort of a metallic bar that came up like a stork's neck, and right at the very tip of it was a sort of metallic-looking ball. Attached to that ball was a kind of a bowl that seemed to just hang off the bottom of the ball, almost like it was magnetically attached to it. He said the whole thing would just sit there, and it would kind of rock and list, almost like a large ship at anchor in a harbor on the ocean, floating on the water. It was literally on a sea of energy.

Dr. [T. Henry] Moray did experiments with different kinds of energy-something that may have been scalar energy – back in the early 1920s or the 1930s, I believe it was. He wrote a book called "The Sea of Energy," and he describes this type of energy.[58] Brad said that when this thing was moving around, the system wasn't energized to its full strength, so components inside the ship were still subject to some influence by gravity. He [said] as it would start to list in one direction, the bowl, because of the influence of gravity, would swing in the same direction. As it started to tilt, it would slide over and it would power up the system on that side, and it would bring it back to a righted position all by itself. Completely unmanned, it would sit there, and it would correct itself just while it was sitting there.

It was all linked fiber-optically. Well, why would that make any difference? Why would you want to have a system that's all linked fiber-optically? The reason is that if you find a way to control gravity, you reduce the mass of it. If you are able to do that, what are the other side benefits? What if you somehow found a way of tapping into this scalar field, this zero-point energy? If what the scientists believe is true, then the zero-point energy is actually what keeps the electrons around the atomic structure of everything in our universe. It keeps them energized-it keeps those little electrons spinning in their different clouds all around the nucleus of every atom in our world. It keeps them going, keeps them from crashing into the nucleus like a satellite orbiting the earth gets pulled into the atmosphere by gravitational drag. Well, if you have a way of interfering with that interaction, that absorption of zero-point energy by those electrons, they begin to slow down.

Every atom in the universe is just like a little gyroscope: it's got all these electrons spinning around the nucleus, and they have a gyroscopic effect, which is the effect we call inertia and mass. We have one nucleus with a proton and a neutron and one electron-hydrogen-spinning around like that: not very much mass, not too much inertia. If you take uranium 235, [with] 235 electrons all spinning around in their different clouds, there is a lot of mass; there is a lot of inertia, because it's like a bigger gyroscope, in a way. At least, that's the analogy that I've kind of picked up here. But, if you have a way of interfering with that absorption of zero-point energy so those electrons become de-energized, they begin to slow down. The effect of that inertia, that gyroscopic effect, begins to drop off; and the

---

[58] See new book, "The Energy Machine of T. Henry Moray" by Moray King, Adventures Unlimited Pub., 2005 – Ed note

mass drops off too, even though the atomic structure is intact; and it's still there— it's still uranium, but it's not as heavy.

One of the things Einstein said was that you could never accelerate anything up to and past the speed of light. If you did, you would have to have all the energy in the universe, because as you accelerate through space, mass increases. One of the old films showing this concept shows a train going faster and faster towards the speed of light, but the train keeps getting bigger and bigger until the engine just can't pull it, so it can never pass the speed of light.

But, what if you have a system, a device, that absorbs that zero-point energy and prevents it from interacting with the atomic structure of the vehicle? And at the same time, it's providing additional power to the capacitor section – this whole electrical system that is going on in the vehicle, that's running. .In effect, the faster you go, the easier it becomes to go up to and exceed the speed of light.

Brad said that in this exhibit at Norton Air Force Base, a tllree star general said that these vehicles were capable of doing light speed or better. Oh, by the way, the largest of these vehicles was about 120 to 130 feet in diameter. I mean, that's massive when you think about it- it's just huge.

There is a scientist in Utah by the name of Moray B. King-he wrote a book called "Tapping the Zero Point Energy." What he maintains is that this energy is embedded in space-time all around us; it's in everything we see. I think it was James Clerk Maxwell who speculated that there's enough of this flux, this electrical charge, in the nothingness of space, that if you could capture all the energy that was embedded in just a cubic yard of space, you'd have enough energy to boil the oceans of the entire world. That's how much energy is sitting there waiting to be tapped. Now, one of the things that Moray B. King said was that the best way to tap that energy is by driving it out of equilibrium. It's just like a bunch of cigarette smoke in a box, but if you somehow send a shockwave through it, you can get force – you can get ripples through it. Then, if you have a way of collecting that energy at the other end, you have a way of tapping into it and using it.

This Alien Reproduction Vehicle, this Flux Liner, has a way of doing that somehow, electronically. Now, Brad had described the fact that this central column has a kind of vacuum chamber in it. The vacuum chamber is one of the things that all of these scientists describe in these over-unity or free energy devices they build. They all have some kind of vacuum tubes, vacuum technology.

Brad maintained that inside this big vacuum chamber in the central column that's inside everything else – inside the flywheel, inside the secondary coils of the Tesla coil, inside the crew compartment – there is mercury vapor. Mercury vapor will conduct electricity, but it produces all kinds of ionic effects. These little molecules of mercury become charged in unusual ways, and if you fire a tremendous amount of electricity through mercury vapor that's in a partial vacuum, there is something special, something unusual that happens in that process.

I believe it's the process that Moray [King] came to describe when he [proposed] driving the energy in the vacuum out of equilibrium, putting some kind of a shockwave through it.

Now, the other thing that I believe happens here, is that as this system begins to tap into this zero-point energy and is drawing it away from the local environment, the whole craft becomes lighter in weight- it becomes partially mass-canceled, if you will, which is one of the reasons why just a little bit of energy in the capacitors could shoot it all over the place.

One of the things that I believe happens, is when you take a system like this and you fire it up, everything in the system starts to become mass-canceled. The next thing that happens is that the electrons that are flowing through the system also become mass-canceled. What does that mean? It means as that system and all the electrons flowing through that big Tesla coil become mass-canceled, it also becomes the perfect super-conductor, which means the efficiency of the systems goes right through the ceiling. You get dramatic efficiency, just like the whole thing was dunked into liquid nitrogen or made out of pure silver or pure gold, which at certain temperatures are perfect conductors – it becomes lighter and can accelerate at incredible speeds.

[The faster it goes, the lighter it gets, and the faster it's able to go. *LW after talking to McCandlish*]

In 1992, I met a man named Kent Sellen and, as it turned out, Kent Sellen and I had a mutual friend: a fellow by the name of Bill Scott, or William Scott, who was a local editor for a trade publication called *Aviation Week and Space Technology*.

Bill Scott used to be a test pilot at Edwards Air Force Base back in the early 1970s, and Kent Sellen had been a crew chief working on the plane that Bill Scott flew. So I was talking to Kent Sellen about this and he nodded his head and smiled a big wide grin, and he winked and he kind of said, "Yes, I know what you're talking about." I [asked] how do you know what I'm talking about? And he [said] "Because I've seen one." At that point, I keyed in on something that John Eppolito of

Intro Vision had told me about something in a hangar – something that someone had seen in a hangar.

So, I [asked] Kent, when I'm meeting him at this air show at Edwards in 1992, was it flat on the bottom and had sloping sides and a dome on the top and little camera things? And he said, "Oh, you've seen one?" I said, let me borrow your pen. I took out a little piece of paper, drew a sketch, and I [asked] does it look like that? He said, "Yes, that's it – that's what it looks like." I [asked] when did you see this? He [replied], "I saw it in 1973." I [asked] where, when did you see it? He [responded], "I was a crew chief, [and] I worked on Bill Scott's plane when he was a test pilot."

He [told me] one night [his] shift supervisor [had told him], "Go out to North Base – they've got a ground power unit for an aircraft that's leaking or failed or something, so we need you to take a tow vehicle out there. Go out, pick it up, bring it back, drop it off at the repair depot; then you can go for the night, because we've finished all our other work." Well, instead of going around the big perimeter road that goes up to the main entrance of North Base, Kent Sellen drove straight across the dry lake bed at Edwards to the North Base facility. He [came] up off the dry lake bed, [rolled] right up on the tarmac, and [was] going down these rows of hangars – they [were] all Quonset-style hangars back then. He [stopped] in front of the first one with the doors cracked, expecting to find this defective ground power unit, and what [did] he see? He [saw] this flying saucer sitting in the hangar, hovering off the ground.

This brings me back to John Eppolito's story about a guy who saw a UFO in a hangar sometime prior to 1982, when I met him. I [asked], what happened? He [said], "This thing was flat on the bottom, [with] sloping sides, little cameras in these little plastic domes all over, [and] there was a door on the side. I wasn't there for 15 seconds, [when] I heard footsteps running up to me, and before I could even turn around and look, there was a machine gun barrel at my throat. A gruff voice [said], "Close your eyes and get on the ground, or we're going to blow your head off."

They put a hood over his head, blindfolded him, hauled him off; and they spent 18 hours debriefing him. While they did, they told him things about this vehicle that my buddy Brad didn't even know.

Brad had said that all of the components in the system were off-the-shelf components-things that you could find right in the inventory. They [had] their own oxygen supply, and he [said] they [ejected] once they [got] below 15,000 feet. The individual seats [came] down off this central column on a set of rails, just like little railroad cars. They

141

[came] off one-by-one, and the parachutes [came] out, and away they [went].

I looked at all this information that I got from Brad, [and] I realized there [was] a mechanical arm that [could] extend out from these little trap doors that [opened] up on the side of the vehicle. It [was] obvious that these things [were] capable of space travel 10 [or] 15 years ago, I was talking to Tom Bearden about scalar effects. One of the things that he said, just off the top of his head, was, "Have you ever wondered why the NASA budget has been cut back so severely? It's because they've got all this other technology that is so much better, so much faster. They are so much better than rocket-propelled spaceships that take months, sometimes years, to get to the outer reaches of the solar system. Why would you put millions and millions of dollars [into] what [amounts to] a public works program for scientists? Why invest all that money when you have this classified system that's used exclusively by the National Security Agency, the CIA, or Air Force Intelligence? It will go anywhere in the solar system in hours, compared to months or years. Why spend all that money on NASA when you've got something that will go there right now?"

When people speculate that there [might] be manned bases on the back side of the moon or there might be bases on Mars, I can tell you that I am almost positive that it's true. In fact, I am positive that it's true.

I have met another man who knows about these things. He said, "I work at the B-2 bomber facility out at Plant 42 in Palmdale and Lancaster. Catty – corner across from the big production facility for the B-2 bomber, down at the southwest corner of the field is the Lockheed Skunkworks – it's that huge complex down there." I said, "Yes, I know exactly where that is." He [said] "In the summer of 1992, I was outside about 10:30 at: night, because I work a late shift and was smoking cigarette, and I noticed that the sheriff's deputies were blocking off all of the streets surrounding Plant 42. They do that anytime there is a classified aircraft coming in to land or is departing from Plant 42."

He [continued], "I noticed all the streets being blocked off, and sitting out in front of this hangar [was] this circular formation of vehicles – but they [were] real weird vehicles. They [were] like a little tractor with a turret on it, and the turret [had] a big mechanical arm with a basket on the end of it. It's the kind of thing that electrical linemen use to work on high-tension power lines, but the baskets [were] all up in the air, and strung from each one of the baskets in this big circle this big black curtain that [came] down, and there [was] a rope that [tied] them a together."

He said, "I looked up above the circular vehicle, and at about 500 feet was this big, black lens-shaped flying saucer, just sitting there above the vehicles. Out of the midst of this group of vehicles [came] a man with a big blue-green handle flashlight. [He shined] it up at the vehicle and [flashed] it three times. There were three blue-green lights underneath the vehicle, and they [flashed back] at him three times.

"Then this thing [lowered] down into [the] cluster of vehicles. The arms all [extended] out over the center and [covered] this craft all up with the curtain – then they all [trundled] into the hangar. The doors [closed], lights [came] on [and] the sheriffs [left]." He said he smoked a lot of cigarettes for the next week after that, waiting to see something else, and a week later his patience was rewarded when the whole process reversed itself. He said that the lights [went] out, the door [opened], [and] this cluster of vehicles [came] out. The arms all [stood] up, and after a while, this thing silently [rose] up to about 500 feet above the vehicles. The guy [came] out with the flashlight, [flashed] three times, [and] it [flashed] its light: at him three times.

Then he said this thing took off, [covering] the entire length of the runway which is right next to the B-2 facility. It went past him and disappeared into the dark in under two seconds – and this vehicle did it without any noise, without any supersonic shockwave, no sonic boom, nothing -- just like it had been fired from canon. He said it changed [his] life. It changed [his] whole perspective, because then he [knew they had] anti-gravity -- massless propulsion.

He [said they had] technology that they might have even recovered from some kind of a spacecraft that came from God-knows-where – some other star system – but, he [said] the fact [was they had] it.

We have found a patent filed by a James King Jr., and this patent looks just like this system except that instead of having a dome for a crew compartment, it has a cylinder in the center. [It has] the same shape, the flat bottom, [and] the sloping sides. It has the coils around the circumference, [and] it has the capacitor plates that are all radially-oriented. This patent was filed initially in 1960 [and] was secured in 1967 – the same year that a photo was taken near Provo, Utah that looks just like [this craft].[59]

The clincher is the guy who filed the patent worked with Townsend Brown. Townsend Brown had worked at a laboratory near Princeton, New Jersey with a scientist by the name of Agnew Bahnson in the

---

[59] See this one in the Patent Section, p. 153   – Ed note

Bahnson Laboratories. They did all these experiments that they were calling electrogravitic propulsion. There is [a] video that was converted from the 16-millimeter film that was shot by Agnew Bahnson's daughter. Originally it was called "Daddy's Laboratory," and it shows all these experiments that Bahnson and Thomas Townsend Brown did, along with their assistant James King (J. Frank King], who filed the patent. That film shows little discs levitating and shooting off sparks and stuff. So, it all kind of comes full circle.[60]

You see that now, they not only have the technology, they've got the technology in deployment. Not only does it fly but it looks just like the patents that were filed back in the 1960s – the same year the photos were taken near Area 51— between Area 51 and Provo, Utah, by a military pilot. .It shows all the same features; it shows all the same shapes. So, the bottom line for me is that regardless of whether you understand all the fine points of the technology, the technology exists and there are people that have seen it. I have seen these things myself: so to me it's just really a matter of time before they bring this technology out of the black and begin to let us use it for other things like pollution-free production of energy. You could probably take a couple of little things that look like those flying saucers and put them around a crank shaft and use them to drive an engine, pollution-free – no use of fuel.

Anyway, the only other thing that I could say, is that when I was talking about the fiber optic control system, that's also one of the things that goes back to the original Roswell account that there were all these little fibers with light going through them, and they couldn't explain what this stuff was. Well, why would you need a fiber optic system in a spaceship? If, suddenly, everything in the vehicle becomes mass-canceled, and even the electrons become mass-canceled, it means that all of the telemetry that's going through your system is going to go haywire. It's going to be like, suddenly, the system goes through a phase change, and everything is super-conducting. So, you have to have some way of maintaining the same level of control for your spark gaps-the control of the amount of electricity that goes out of the capacitors-so that when you change the control stick, you still get the same amount of movement and deflection in the system, even when you go into a state of mass-cancellation or partial mass-cancellation, because the electrons are also mass-canceled, so they become super-conducting circuits.

---

[60] This "Brown-Bahnson Lab" silent video #606 is available from Integrity Research Institute. See Publications Section. – Ed note

Why use fiber optics? Because photons have no mass, so they are unaffected. That means any information, any telemetry that you send back and forth to your computer gets there. It doesn't matter if the computer functions at the super-conducting level, because it just makes it faster, more efficient, smarter. You want to be able to control the aircraft so it doesn't crash, and what's the best way to do it? With fiber optics.

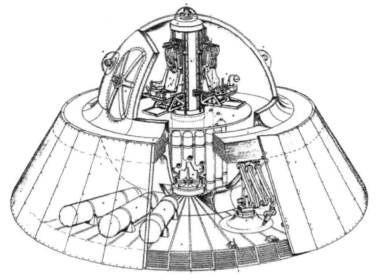

Electrogravitic Craft Demonstration
Norton Air Force Base 1988
Copyright © 2000, Mark McCandlish

PATENT SECTION

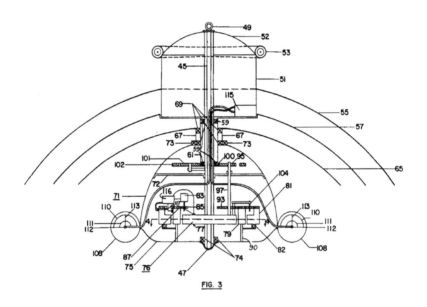

FIG. 3

A.H. Bahnson's "Electrical Thrust Producing Device,"
Figure 2 embodiment with electric field gradient
shaping electrodes from his US Patent #2,958,790,
Nov. 1, 1960

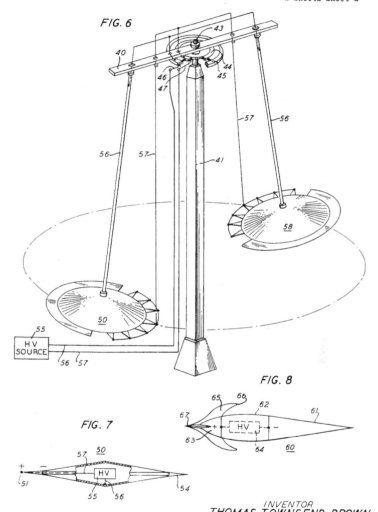

FIG. 6

FIG. 8

FIG. 7

INVENTOR
THOMAS TOWNSEND BROWN

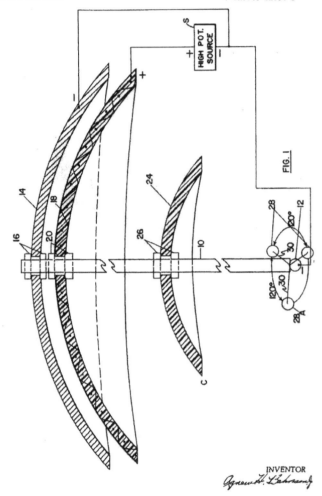

FIG. 1

INVENTOR

Agnew H. Bahnson Jr

BY

Watson Cole, Grindle & Watson
ATTORNEYS

148

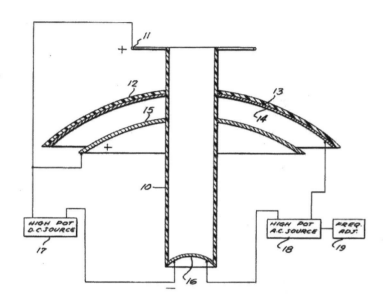

INVENTOR
AGNEW H.BAHNSON,JR, DECEASED
by- WACHOVIA BANK & TRUST CO.,
EXECUTOR
BY
Mason, Fenwick & Lawrence
ATTORNEYS

149

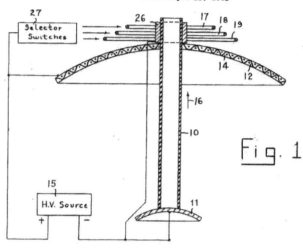

Fig. 1

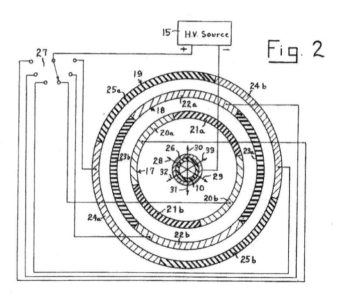

Fig. 2

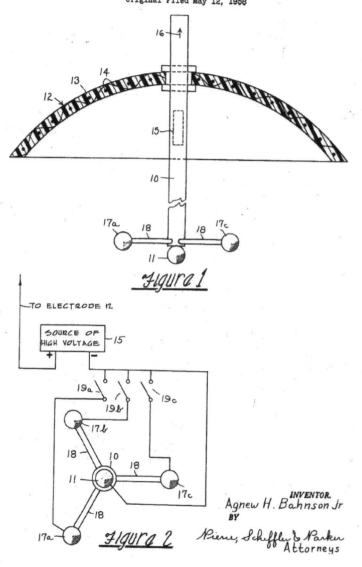

Figure 1

TO ELECTRODE 12

SOURCE OF HIGH VOLTAGE — 15

+ —

Figure 2

INVENTOR.

Agnew H. Bahnson Jr

BY

Pierce, Schiffler & Parker

Attorneys

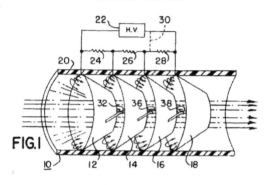

Aug. 23, 1966 T. T. BROWN 3,267,860

ELECTROHYDRODYNAMIC FLUID PUMP

Filed Dec. 31, 1964

H.V.
FIG.1
FIG.2
FIG.3
FIG.4
FIG.6

FIG.5

INVENTOR
THOMAS TOWNSEND BROWN

BY

ATTORNEY

All of the rest of T.T. Brown's patents can be found in the first *Electrogravitics* volume. – Ed note

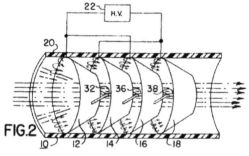

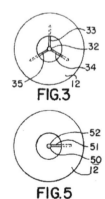

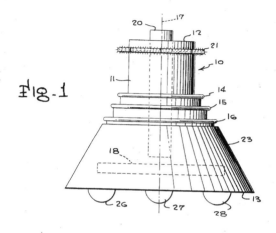

Fig-1

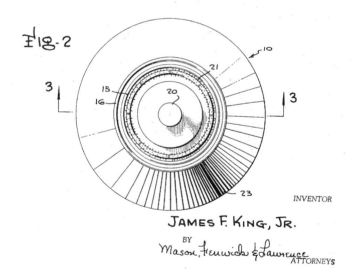

Fig-2

INVENTOR

JAMES F. KING, JR.

BY

Mason, Fenwick & Lawrence

ATTORNEYS

Ring #21 is a high voltage "sector control" ring that exerts an electric field toward a 3-phase set of rings #14, 15, 16 with capacitors for controlling the phase angle and attitude, as seen in Figs. 3, 4.   - Ed note

153

(12) **United States Patent** (10) **Patent No.:** **US 6,775,123 B1**
Campbell (45) **Date of Patent:** **Aug. 10, 2004**

(54) **CYLINDRICAL ASYMMETRICAL CAPACITOR DEVICES FOR SPACE APPLICATIONS**

(75) Inventor: **Jonathan W. Campbell**, Harvest, AL (US)

(73) Assignee: **The United States of America as represented by the Administrator of the National Aeronautics and Space Administration**, Washington, DC (US)

(*) Notice: Subject to any disclaimer, the term of this patent is extended or adjusted under 35 U.S.C. 154(b) by 0 days.

(21) Appl. No.: **10/446,282**

(22) Filed: **May 27, 2003**

(51) **Int. Cl.⁷** **H01G 4/228**
(52) **U.S. Cl.** **361/306.1; 361/811**
(58) **Field of Search** 361/306.1, 311, 361/15, 16, 17, 715, 821, 811

(56) **References Cited**

U.S. PATENT DOCUMENTS

6,317,310 B1 * 11/2001 Campbell 361/306.1

6,411,493 B2 * 6/2002 Campbell 361/306.1

* cited by examiner

Primary Examiner—Anthony Dinkins
(74) *Attorney, Agent, or Firm*—James J. McGroary; Ross R. Hunt, Jr.

(57) **ABSTRACT**

An asymmetrical capacitor system is provided which creates a thrust force. The system is adapted for use in space applications and includes a capacitor device provided with a first conductive element and a second conductive element axially spaced from the first conductive element and of smaller axial extent. A shroud supplied with gas surrounds the capacitor device. The second conductive element can be a wire ring or mesh mounted on dielectric support posts affixed to a dielectric member which separates the conductive elements or a wire or mesh annulus surrounding a barrel-shaped dielectric member on which the first element is also mounted. A high voltage source is connected across the conductive elements and applies a high voltage to the conductive elements of sufficient value to create a thrust force on the system inducing movement thereof.

30 Claims, 3 Drawing Sheets

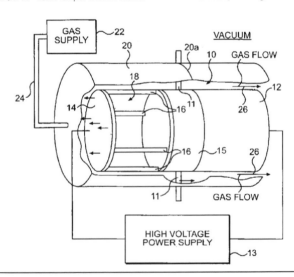

Also see #6,411, 493 and the first patent in this NASA series, US #6,317,310, which caused protest letters to the Patent Office for not referencing any T.T. Brown or Agnew Bahnson patents, nor Hector Serrano's invention.   – Ed note

PCT

WORLD INTELLECTUAL PROPERTY ORGANIZATION
International Bureau

INTERNATIONAL APPLICATION PUBLISHED UNDER THE PATENT COOPERATION TREATY (PCT)

(51) International Patent Classification 7 : F03H	A2	(11) International Publication Number: **WO 00/58623**
		(43) International Publication Date: 5 October 2000 (05.10.00)

(21) International Application Number: PCT/US00/05667

(22) International Filing Date: 3 March 2000 (03.03.00)

(30) Priority Data:
60/123,086   5 March 1999 (05.03.99)   US

(71) Applicant: GRAVITEC, INC. [US/US]; P.O. Box 421928, 808 North Main Street, Kissimmee, FL 34742 (US).

(72) Inventor: SERRANO, Hector, L.; 1110 Catherine Street, Kissimmee, FL 34741 (US).

(74) Agent: NAPOLITANO, Carl, M.; Allen, Dyer, Doppelt, Milbrath & Gilchrist, P.A., Suite 1401, 255 South Orange Avenue, P.O. Box 3791, Orlando, FL 32801–3791 (US).

(81) Designated States: AE, AL, AM, AT, AU, AZ, BA, BB, BG, BR, BY, CA, CH, CN, CR, CU, CZ, DE, DK, DM, EE, ES, FI, GB, GD, GE, GH, GM, HR, HU, ID, IL, IN, IS, JP, KE, KG, KP, KR, KZ, LC, LK, LR, LS, LT, LU, LV, MA, MD, MG, MK, MN, MW, MX, NO, NZ, PL, PT, RO, RU, SD, SE, SG, SI, SK, SL, TJ, TM, TR, TT, TZ, UA, UG, UZ, VN, YU, ZA, ZW, ARIPO patent (GH, GM, KE, LS, MW, SD, SL, SZ, TZ, UG, ZW), Eurasian patent (AM, AZ, BY, KG, KZ, MD, RU, TJ, TM), European patent (AT, BE, CH, CY, DE, DK, ES, FI, FR, GB, GR, IE, IT, LU, MC, NL, PT, SE), OAPI patent (BF, BJ, CF, CG, CI, CM, GA, GN, GW, ML, MR, NE, SN, TD, TG).

Published
Without international search report and to be republished upon receipt of that report.

(54) Title: PROPULSION DEVICE AND METHOD EMPLOYING ELECTRIC FIELDS FOR PRODUCING THRUST

(57) Abstract

Thrust is provided to a vehicle (12) using a self–contained device (10) for producing the thrust through a preselected shaping of an electric field. The device (10) includes a core (28) carried by a housing (18), with both the core (28) and the housing (18) formed from a material having a high dielectric constant. A plurality of cells (22) are carried by the housing (18) and formed around the core (28), with each cell (22) having a high dielectric (36) sandwiched between an electrode (38) and a lower dielectric (40). Multiple plates (26) are stacked along a longitudinal axis (24) of the core (28) with the electric wire (46) carried through the high dielectric (36) for connection with the electrodes (38) of each plate (26). Positive and negative voltage is provided to adjacent plates (26) at a rapidly changing rate to provide thrust resulting from non–linear electric field paths created through the device (10) as a result of the cell (22) and surrounding material (42, 28) configuration.

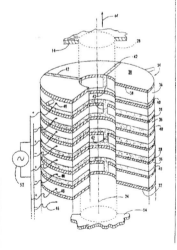

The "rapidly changing rate" of this voltage is reminiscent of the Air Force electrogravitic craft seen on p. 145. In other systems, like bioelectromagnetics, when a static electric field exhibits an effect, pulsing the EMF usually increases or amplifies the effect. – Ed note

155

US006098924A

United States Patent [19]

Woodward et al.

[11] Patent Number: 6,098,924

[45] Date of Patent: Aug. 8, 2000

[54] **METHOD AND APPARATUS FOR GENERATING PROPULSIVE FORCES WITHOUT THE EJECTION OF PROPELLANT**

[75] Inventors: **James Woodward**, Anaheim; **Thomas Mahood**, Irvine, both of Calif.

[73] Assignee: **California State University, Fullerton Foundation**

[21] Appl. No.: **09/236,188**

[22] Filed: **Jan. 23, 1999**

[51] Int. Cl.^{7} **B64D 35/00**
[52] U.S. Cl. **244/62**; 244/172
[58] Field of Search .... 244/62, 53 R, 244/172

[56] **References Cited**

U.S. PATENT DOCUMENTS

5,280,864   1/1994   Woodward 244/62

Primary Examiner—Galen L. Barefoot

Attorney, Agent, or Firm—Leonard Tachner

[57] **ABSTRACT**

Mach's principle and local Lorentz-invariance together yield the prediction of transient rest mass fluctuations in accelerated objects. These restmass fluctuations, in both principle and practice, can be quite large and, in principle at least, negative. They suggest that exotic space time transport devices may be feasible, the least exotic being "impulse engines", devices that can produce accelerations without ejecting any material exhaust. Such "impulse engines" rely on inducing transient mass fluctuations in conventional electrical circuit components and combining them with a mechanically coupled pulsed thrust to produce propulsive forces without the ejection of any propellant. The invention comprises a method of producing propellant-less thrust by using force transducers (piezoelectric devices or their magnetic equivalents) attached to resonant mechanical structures. The force transducers are driven by two phase-locked voltage waveforms so that the transient mass fluctuation and mechanical excursion needed to produce a stationary thrust are both produced in the transducer itself.

20 Claims, 19 Drawing Sheets

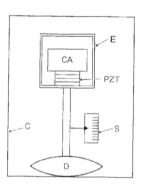

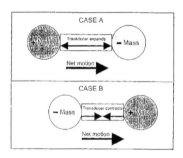

(12) **United States Patent**
Woodward et al.

(10) Patent No.: **US 6,347,766 B1**
(45) **Date of Patent:** Feb. 19, 2002

(54) **METHOD AND APPARATUS FOR GENERATING PROPULSIVE FORCES WITHOUT THE EJECTION OF PROPELLANT**

(76) Inventors: **James Woodward**, 5549 Stetson Ct., Anaheim, CA (US) 92807; **Thomas Mahood**, 4 La Paloma Dr., Irvine, CA (US) 92620

(*) Notice: Subject to any disclaimer, the term of this patent is extended or adjusted under 35 U.S.C. 154(b) by 0 days.

(21) Appl. No.: **09/549,475**

(22) Filed: **Apr. 14, 2000**

Related U.S. Application Data

(63) Continuation-in-part of application No. 09/236,188, filed on Jan. 23, 1999, now Pat. No. 6,098,924.

(51) **Int. Cl.**[7] **B04D 35/00**
(52) **U.S. Cl.** **244/62**; 244/172
(58) **Field of Search** .... 244/62, 172, 158 R, 244/53 R; 60/203.1

(56) **References Cited**

U.S. PATENT DOCUMENTS

5,280,864 A * 1/1994 Woodward 244/62
6,098,924 A * 8/2000 Woodward et al. 244/62

OTHER PUBLICATIONS

Woodward, "A New Experimental approach to Mach's principle and relativistic graviation" Foundations of physics letters, vol. 3, No. 5 1990.*

* cited by examiner

Primary Examiner—Galen L. Barefoot
(74) *Attorney, Agent, or Firm*—Leonard Tachner

(57) **ABSTRACT**

Mach's principle and local Lorentz-invariance together yield the prediction of transient rest mass fluctuations in accelerated objects. These restmass fluctuations, in both principle and practice, can be quite large and, in principle at least, negative. They suggest that exotic space time transport devices may be feasible, the least exotic being "impulse engines", devices that can produce accelerations without ejecting any material exhaust. Such "impulse engines" rely on inducing transient mass fluctuations in conventional electrical circuit components and combining them with a mechanically coupled pulsed thrust to produce propulsive forces without the ejection of any propellant. The invention comprises a method of producing propellant-less thrust by using force transducers (piezoelectric devices or their magnetic equivalents) attached to resonant mechanical structures. The force transducers are driven by two (or more) phase-locked voltage waveforms so that the transient mass fluctuation and mechanical excursion needed to produce a stationary thrust are both produced in the transducer itself.

22 Claims, 20 Drawing Sheets

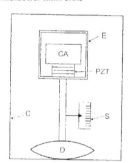

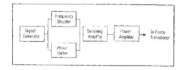

US005142861A

United States Patent [19]

Schlicher et al.

[11] **Patent Number:** **5,142,861**

[45] **Date of Patent:** **Sep. 1, 1992**

[54] **NONLINEAR ELECTROMAGNETIC PROPULSION SYSTEM AND METHOD**

[76] Inventors: Rex L. Schlicher, 8230 Stationhouse Ct., Lorton, Va. 22079-1204; Steven M. Rinaldi, ODC/AFSC, American Embassy, E-401 APO New York, N.Y. 09777-5000; David J. Hall, 460 Tahoe Dr., Pittsburgh, Pa. 15239; Peter M. Ranon, P.O. Box 6074, San Pedro, Calif. 90734; Charles E. Davis, 13400 Lomas Blvd., N.E. #221, Albuquerque, N. Mex. 87112

[21] Appl. No.: 691,889

[22] Filed: Apr. 26, 1991

[51] Int. Cl.⁵ F02K 11/00
[52] U.S. Cl. 60/203.1; 343/896
[58] Field of Search 60/200.1, 203.1, 204; 315/34; 343/867, 896

[56] **References Cited**

U.S. PATENT DOCUMENTS

2,617,033	11/1952	Posthumus	343/867
3,229,297	1/1966	Bell et al.	343/896
3,366,962	1/1968	Kulik et al.	343/896

FOREIGN PATENT DOCUMENTS

139151	12/1979	German Democratic Rep.	60/200.1
8900073	8/1990	Netherlands	60/200.1

Primary Examiner—Louis J. Casaregola

Attorney, Agent, or Firm—Bernard E. Franz; Donald J. Singer

[57] **ABSTRACT**

An electromagnetic propulsion system based on an extremely low frequency (elf) radiating antenna structure driven by a matched high current pulsed power supply is described. The elf antenna structure resembles a modified three dimensional multiple-turn loop antenna whose geometry is optimized for the production of reaction thrust rather than the radiation of electromagnetic energy into space. The antenna structure is current driven rather than voltage drive. Rigid three dimensional geometric asymmetry, made up of flat electrical conductors that form a partially closed volume in the loop antenna structure, trap magnetic flux thereby causing a magnetic field density gradient along a single axis. This magnetic field density gradient then causes an imbalance in the magneto- mechanical forces that normally result from the interactions of the loop antenna's internal magnetic field with the current in the conductors of the loop antenna structure, as described by the Lorentz Force Law. The pulsed power supply is designed to provide the proper waveform to the antenna structure at an impedance matching the load impedance of the antenna. The rise time and shape of the input current waveform is crucial to maximizing the production of reaction thrust. Input voltage is at a nominal value sufficient to allow the desired high input current.

14 Claims, 10 Drawing Sheets

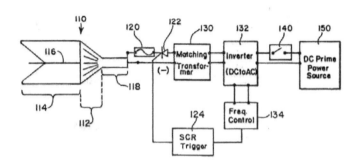

US005197279A

United States Patent [19]

Taylor

[11] Patent Number: 5,197,279

[45] Date of Patent: Mar. 30, 1993

[54] **ELECTROMAGNETIC ENERGY PROPULSION ENGINE**

[76] Inventor: **James R. Taylor**, 1907 May Cir., Fultondale, Ala. 35068

[21] Appl. No.: **847,684**

[22] Filed: **Mar. 6, 1992**

Related U.S. Application Data

[63] Continuation-in-part of Ser. No. 459,441, Jan. 2, 1990, abandoned.

[51] Int. Cl.⁵ F03H 5/00
[52] U.S. Cl. 60/203.1; 60/200.1
[58] Field of Search 60/200.1, 201, 202, 60/203.1

[56] **References Cited**

FOREIGN PATENT DOCUMENTS

1586195 2/1970 France .
2036646 12/1970 France .
58-32976(A) 2/1983 Japan .

OTHER PUBLICATIONS

Megagauss Fields, by J. G. Linhart, Physics Today, Feb. 1966, pp. 37–42.
Static and Dynamic Electricity, by W. R. Smythe, McGraw-Hill Book Company, Inc., New York, New York, 1950, pp. 447 and 448.
Principles of Electricity and Electromagnetism, by G. P. Harnwell, 2nd Edition, McGraw-Hill Book Company, Inc., New York, New York, 1949, pp. 572–579.
Introduction to Modern Physics, by Richtmyer and Kennard, 4th Edition, McGraw-Hill Book Company, Inc., New York, New York, 1947, pp. 58–61 and 146–149.
Electromagnetic Fields, Energy and Forces, by Fano, Chu and Adler, John Wiley and Sons, Inc., New York, New York, pp. 421–425.
The Feynman Lectures on Physics, by Feynman, Leighton & Sands, Addison-Wesley Publishing Co, New York, N.Y., pp. 17-5 to 17-6, 27-9 to 27-11, 34-10 to 34-11.
McGraw-Hill Encyclopedia of Science and Technology,

McGraw-Hill Book Co. Inc., New York, New York, 1977, vol. 8, pp. 626–629.
Superconducting Magnets, by Martin Wilson, published by Oxford University Press, second edition, 1989, p. 3.
Materials and Techniques for Electron Tubes, by Walter H. Kohl, Reinhold Publishing Corporation, NY, NY, 1962, pp. 92, 93, 109, 114, 115.

Primary Examiner—Louis J. Casaregola
Attorney, Agent, or Firm—Jacobson, Price, Holman & Stern

[57] **ABSTRACT**

An electromagnetic energy propulsion engine system including a hollow housing having a front part (50) and a rear end part (4) of material transparent to the passage of electromagnetic fields, electromagnetic field generating solenoidal windings (23), (25), having central axes parallel with the central axis of the engine and axially spaced from each other to provide a forward field generating winding (25) and a rear field generating winding (23), a power source (44), a control computer (42), and a power pulse generator (40) connected between the electromagnetic field generating windings and the power source and control computer. The forward field generating winding generates a rearwardly directed magnetic field toward the rear wall parallel to the central axis, and the rear field generating winding produces a forwardly directed magnetic field opposing the rearwardly directed magnetic field of the forward field generating winding so that the rearwardly directed magnetic field repels forwardly directed pulses of the rear magnetic generating winding. As the electrical current conduction in the rear field generating winding suddenly reduces, the continuing rearwardly directed magnetic field force transmits pulsating magnetic field energy produced by the rear field generating winding through the rear of the housing. The reaction to the rearwardly transmitted field energy produces a thrust propelling the engine and a vehicle in which it is mounted.

26 Claims, 20 Drawing Sheets

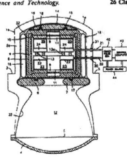

PUBLICATIONS - Information on Electrogravitics

T. T. Brown's Electrogravitics Research by Thomas Valone, PhD, PE. Articles on electrogravitics with diagrams and references, the Townsend Brown Notebooks, repeated Brown experiment, and the Bahnson Lab film/video with still photos.   #603 28 pg. $5

The Townsend Brown Electro-Gravity Device. A comprehensive evaluation by the Office of Naval Research, with accompanying documents, issued Sept. 15, 1952. Every page was marked "confidential" until cancelled by ONR. #612 22 pg. $6

Thomas Townsend Brown: Bahnson Lab 1958-1960. This rare find is a documentary revealing short segments of all Brown-Bahnson lab experiments. #606   1 hr silent NTSC VHS videotape $20

SubQuantum Kinetics: The Alchemy of Creation. by Dr. Paul LaViolette. Predicts electrogravitic forces. #605 268 pg. book $25

Observations of a Massive Torque Pendulum: Gravity Measurements During an Eclipse by Erwin Saxl and Mildred Allen, Well documented experiment with articles #702   52 pages $10

The Zinsser Effect by Thomas Valone, Complete reports on a German electrogravitic invention and analysis with photos, graphs, and patent, mostly in English. #701   130 pages $20

Electrogravitics Reference List by Robert Stirniman. Extensive listing of papers, abstracts, websites, and book title/descriptions. Good aid for research in the field.   #616   55 pages $15

ORDERING INFORMATION: Shipping cost is $5 for U.S. residents. Add $10 for Canada and $15 for overseas. Integrity Research Institute, 5020 Sunnyside Avenue, Suite 209, Beltsville MD 20705. Phone: 800-295-7674 or 202-452-7674. Visit **www.IntegrityResearchInstitute.org** to order online.

NOTE: You are invited to email IRI@erols.com or mail a request for a "**Free EGII CD**" in the subject with your postal mailing address. The computer data CD (for Windows) contains an electronic version of both Volume I and II, as well as a narrated slide show and much more. Since it is labor intensive to insert them in every book, this way we are happy to send them at no cost only to those who really want the added information. ☺